D0266875

IN 25 VOLUMES
*Each title on a specific drug or drug-related problem*

| | |
|---|---|
| ALCOHOL | ***Alcohol*** *and Alcoholism* |
| | ***Alcohol*** *Customs and Rituals* |
| | ***Alcohol*** *Teenage Drinking* |
| HALLUCINOGENS | ***Flowering Plants*** *Magic in Bloom* |
| | ***LSD*** *Visions or Nightmares* |
| | ***Marijuana*** *Its Effects on Mind & Body* |
| | ***Mushrooms*** *Psychedelic Fungi* |
| | ***PCP*** *The Dangerous Angel* |
| NARCOTICS | ***Heroin*** *The Street Narcotic* |
| | ***Methadone*** *Treatment for Addiction* |
| | ***Prescription Narcotics*** *The Addictive Painkillers* |
| NON-PRESCRIPTION DRUGS | ***Over-the-Counter Drugs*** *Harmless or Hazardous?* |
| SEDATIVE HYPNOTICS | ***Barbiturates*** *Sleeping Potion or Intoxicant* |
| | ***Inhalants*** *The Toxic Fumes* |
| | ***Methaqualone*** *The Quest for Oblivion* |
| | ***Tranquillizers*** *The Cost of Calmness* |
| STIMULANTS | ***Amphetamines*** *Danger in the Fast Lane* |
| | ***Caffeine*** *The Most Popular Stimulant* |
| | ***Cocaine*** *A New Epidemic* |
| | ***Nicotine*** *An Old-Fashioned Addiction* |
| UNDERSTANDING DRUGS | ***The Addictive Personality*** |
| | ***Escape from Anxiety and Stress*** |
| | ***Getting Help*** *Treatments for Drug Abuse* |
| | ***Treating Mental Illness*** |
| | ***Teenage Depression and Drugs*** |

# FLOWERING PLANTS

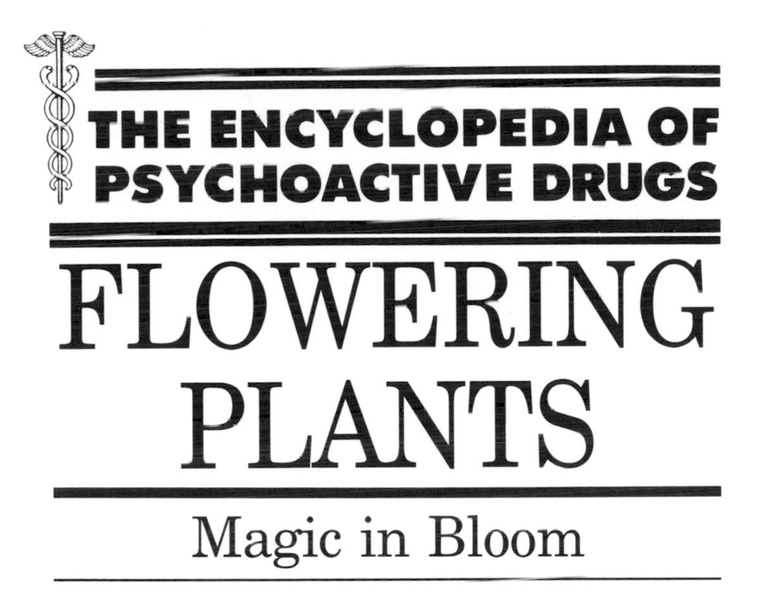

THE ENCYCLOPEDIA OF PSYCHOACTIVE DRUGS

# FLOWERING PLANTS

## Magic in Bloom

P. MICK RICHARDSON, Ph.D.
*New York Botanical Garden*

GENERAL EDITOR (U.S.A.)
Professor Solomon H. Snyder, M.D.
*Distinguished Service Professor of Neuroscience, Pharmacology and Psychiatry at*
*The Johns Hopkins University School of Medicine*

GENERAL EDITOR (U.K.)
Professor Malcolm H. Lader, D.Sc., Ph.D., M.D., F.R.C. Psych.
*Professor of Clinical Psychopharmacology*
*at the Institute of Psychiatry, University of London,*
*and Honorary Consultant to the Bethlem Royal and Maudsley Hospitals*

**Burke Publishing Company Limited**
LONDON TORONTO NEW YORK

First published in the United States of America 1986

This edition first published 1988
New material contained in this edition


**Acknowledgements**
The publishers are grateful to those individuals and organizations, including HM Customs and Excise, for permission given to use material and illustrations included in this publication.

**CIP data**
Richarson, P. Mark
Flowering Plants – (Encyclopedia of psychoactive)
1. Narcotic Plants
I. Title II. Series.
583'.04634
ISBN 0 222 01449 0 Hardbound
ISBN 0 222 01450 4 Paperback

Burke Publishing Company Limited
Pegasus House, 116-120 Golden Lane, London EC1Y 0TL, England.
Printed in Spain by Jerez Industrial, S.A.

# CONTENTS

UPI/BETTMANN NEWSPHOTOS

*In many primitive societies, the shaman, or medicine man, ingests hallucinogenic plants, inducing a trance under which he performs ritual actions to cure disease and foretell the future.*

# INTRODUCTION

The late twentieth century has seen the rapid growth of both the legitimate medical use and the illicit, non-medical abuse of an increasing number of drugs which affect the mind. Both use and abuse are very high in general in the United States of America and great concern is voiced there. Other Western countries are not far behind and cannot afford to ignore the matter or to shrug off the consequent problems. Nevertheless, differences between countries may be marked and significant: they reflect such factors as social habits, economical status, attitude towards the young and towards drugs, and the ways in which health care is provided and laws are enacted and enforced.

Drug abuse particularly concerns the young but other age groups are not immune. Alcoholism in middle-aged men and increasingly in middle-aged women is one example, tranquillizers in women another. Even the old may become alcoholic or dependent on their barbiturates. And the most widespread form of addiction, and the one with the most dire consequences to health, is cigarette-smoking.

Why do so many drug problems start in the teenage and even pre-teenage years? These years are critical in the human life-cycle as they involve maturation from child to adult. During these relatively few years, adolescents face the difficult task of equipping themselves physically and intellectually for adulthood and of establishing goals that make adult life worthwhile while coping with the search for personal identity, assuming their sexual roles and learning to come to terms with authority. During this intense period of growth and activity, bewilderment and conflict are inevitable, and peer pressure to experiment and to escape from life's apparent problems becomes overwhelming. Drugs are increasingly available and offer a tempting respite.

Unfortunately, the consequences may be serious. But the penalties for drug-taking must be put into perspective. Thus,

*An illustration from an Incan manuscript depicts two people cultivating medicinal herbs. Flowering plants have been used in some parts of the world for ceremonial and medicinal purposes since antiquity.*

addicts die from heroin addiction but people also die from alcoholism and even more from smoking-related diseases. Also, one must separate the direct effects of drug-taking from those indirectly related to the life-style of so many addicts. The problems of most addicts include many factors other than drug-taking itself. The chaotic existence or social deterioration of some may be the cause rather than the effect of drug abuse.

Drug use and abuse must be set into its social context. It reflects a complex interaction between the drug substance (naturally-occurring or synthetic), the person (psychologically normal or abnormal), and society (vigorous or sick). Fads effect drug-taking, as with most other human activities, with drugs being heavily abused one year and unfashionable the next. Such swings also typify society's response to drug abuse. Opiates were readily available in European pharmacies in the last century but are stringently controlled now. Marijuana is accepted and alcohol forbidden in many Islamic countries; the reverse obtains in most Western countries.

The use of psychoactive drugs dates back to prehistory. Opium was used in Ancient Egypt to alleviate pain and its main constituent, morphine, remains a favoured drug for pain relief. Alcohol was incorporated into religious ceremonies in the cradles of civilization in the Near and Middle East and has been a focus of social activity ever since. Coca leaf has been chewed by the Andean Indians to lessen fatigue; and its modern derivative, cocaine, was used as a local anaesthetic. More recently, a succession of psychoactive drugs have been synthesized, developed and introduced into medicine to allay psychological distress and to treat psychiatric illness. But, even so, these innovations may present unexpected problems, such as the difficulties in stopping the long-term use of tranquillizers or slimming-pills, even when taken under medical supervision.

*The Encyclopedia of Psychoactive Drugs* provides information about the nature of the effects on mind and body of alcohol and drugs and the possible results of abuse. Topics include where the drugs come from, how they are made, how they affect the body and how the body deals with these chemicals; the effects on the mind, thinking, emotions, the will and the intellect are detailed; the processes of use and

*An illustration of a* Brugmansia *plant, a member of the potato family.* Brugmansia *plants do not occur in the wild but are cultivated in South America by various Indian tribes, who use the plants for their medicinal and hallucinatory properties.*

abuse are discussed, as are the consequences for everyday activities such as school work, employment, driving, and dealing with other people. Pointers to identifying drug users and to ways of helping them are provided. In particular, this series aims to dispel myths about drug-taking and to present the facts as objectively as possible without all the emotional distortion and obscurity which surrounds the subject. We seek neither to exaggerate nor to play down the complex topics concerning various forms of drug-abuse. We hope that young people will find answers to their questions and that others—parents and teachers, for example—will also find the series helpful.

The series was originally written for American readers by American experts. Often the problem with a drug is particularly pressing in the USA or even largely confined to that country. We have invited a series of British experts to adapt the series for use in non-American English-speaking countries and believe that this widening of scope has successfully increased the relevance of these books to take account of the international drug scene.

This volume, originally written by P. Mick Richardson and adapted here by Malcolm H. Lader, deals with the fascinating topic of naturally occurring hallucinogens. These are substances termed "alkaloids", which are found naturally in a wide variety of plant products from exotic tropical flowers such as morning glory and the Barbados cherry to the nutmeg and humble potato. When taken by mouth, smoked, snuffed or rubbed on the skin, a large number of psychological effects can follow, of which the most dramatic are hallucinations, subjective experiences of an often bizarre nature.

Discovery of the mind-altering properties of such plants goes back thousands of years. In this time, native societies integrated the plant substances into complex social, religious and mystical practices. Recently, knowledge about many of these substances has become more generally available leading to experimentation in other societies. Sometimes severe reactions or even deaths occur. This book sets out what is known about these natural hallucinogens and emphasizes their potential dangers.

**M. H. Lader**

DR. RICHARD E. SCHULTES

*A medicine man from the Brazilian Amazon area lights a torch at the beginning of an all-night ceremony that includes the drinking of* caapi, *a hallucinogenic drink made from the bark of* Banisteriopsis caapi.

# AUTHOR'S PREFACE

There are between 250,000 and 350,000 species of flowering plants, and it is likely that over the course of thousands of years people have tasted most of them. Even today, members of "primitive" cultures use a high percentage of the plants native to their village. Some of these plants have medicinal value, some are poisonous, and others produce altered states of awareness. The plants that cause these altered states are said to be *hallucinogenic* because when they are consumed they produce hallucinations, or sensory impressions that have no basis in the physical world.

Most of the known hallucinogenic plants are found in the New World, mainly in the tropics and subtropics. During the past three decades the role of these plants in the lives of people in these regions has come to the attention of Western scientists. It is likely, however, that some plants and their uses remain unknown. A more vigorous search in both the Old World and the New World will undoubtedly uncover other hallucinogens.

## *Early Use of Hallucinogenic Plants*

It is impossible to determine when the people of primitive societies first consumed hallucinogenic plants. Scientists do know that 50,000 years ago Neanderthal cultures used herbal remedies, but no remains of hallucinogenic plants have been discovered in archaeological sites dating back to this period. The earlist hallucinogens associated with human remains are mescal beans (the seeds of *Sophora secundiflora*), and peyote cacti (*Lophophora williamsii*) (see Chapter 4). These hallucinogenic seeds and cacti have been found in caves in Texas and in adjacent parts of Mexico, where they were used by the occupants as early as 10,000 years ago. Mescal beans were used continuously until the early part of the 20th century, when they were replaced in Indian ceremonies by the peyote cactus, a less dangerous plant.

Hallucinogens, such as peyote and mescal beans, that are found in arid regions decompose slowly enough to become preserved. Thus, their remains can be found thousands of years later amidst other archaeological artifacts. But it is more difficult to date the earliest use of other hallucinogenic plants, because many of these plants grow in tropical areas, where plant remains decompose very rapidly. Therefore, they are seldom found at archaeological sites.

The next earliest known use of hallucinogenic plants comes from indirect evidence in the Old World. Fibres of hemp from the cannabis plant (of which marijuana is one type) have been found at sites in China dated 4000 BC and at sites in Turkestan in the Soviet Union dated 3000 BC. Although it is not possible to determine if the early inhabitants of these areas consumed the cannabis to experience its hallucinogenic effects, Chinese writings from 2000 BC suggest that the plant's hallucinogenic properties were known at this time. Cannabis has been used continuously from these early dates up to the present. Songs from India from as early as 600 BC contain reference to an intoxicating resin, presumably from cannabis. Excavations of Scythian tombs in Asia dated 400 BC revealed braziers (pans for holding burning coals) and cannabis leaves and seeds. Scientists believe that people gather around the braziers, threw the seeds onto the hot stones, and inhaled the fumes (see chapter 3).

Early manuscripts also describe the use of hallucinogenic plants in the potato family. For example, the Egyptians mention henbane (*Hyoscyamus niger*), a member of the nightshade family, as early as 1500 BC. Ancient Greek writings describe several uses of henbane, including its purported ability to give the user the power to foretell the future. Numerous medieval encyclopedias of medicinal plants describe the folklore and uses of henbane, mandrake, and the deadly nightshade. The nightly "trips" taken by witches who had ingested these plants are also described in many European texts dating from the Middle Ages.

The dates of the earliest use of most other hallucinogens will never be known with any degree of certainty. European explorers, seeking new trade routes and new lands to colonize, were frequently the first people from the West to come into contact with hallucinogenic plants. Missionaries, who followed the explorers into the new lands, inevitably tried to stamp out the local religions—many of which included the use of hallucinogens—and replace them with Christianity. They often kept detailed records, and in some of these there are descriptions of the use of many plants by the native inhabitants. In some cases, the natives were able to successfully resist the missionaries' influence, and thus some non-Christian religions did survive. Examples of such religions are the Bwiti and Mbiri cults in West Africa which use the hallucinogenic iboga plant, *Tabernanthe iboga.*

Unfortunately, a large percentage of the knowledge of hallucinogenic plants was lost in early colonial times, and many of these plants have become known to scientists only in the last 30 years. Other uses of hallucinogens may have been

***A 19th-century engraving entitled "Elliot, the First Missionary among the Indians". Missionaries tried to replace the New World's religions, many of which included the use of hallucinogens, with Christianity.***

lost forever when native populations were exterminated by murder, disease, or enslavement—the fate of many Indians of South America.

The use of hallucinogens in primitive societies, especially those in the New World, has been extensively studied by Professor Richard Evans Schultes of Harvard University. These societies restrict the use of hallucinogens to magic, medical, and religious purposes. Sometimes the hallucinogens are used only by a shaman, or medicine man, but other times, such as in some religious ceremonies, all adult males use the drugs. Children are rarely allowed to consume hallucinogenic plants, though in some societies puberty rites include the child's first use of these plants. Some of these initiation rites cause the adolescent great physical discomfort and may include fierce whippings and the inducement of nausea and vomiting. Frequently, the adolescents become deranged, a condition which the adults interpret as an indication that childhood is being forgotten and adulthood has commenced. Such ceremonies occur in both the Old World and the New World and involve plants from the Solanaceae, Apocynaceae, and Malpighiaceae families.

## *Present-Day Use of Hallucinogens*

Primitive people incorporated hallucinogenic plants into their culture and thus severely restricted their use. Within this context, the powerful drugs in the plants caused little or no harm. These people had hundreds of years to gain an understanding of the interaction between the human body and hallucinogens.

A different situation exists in the modern world, where use of these drugs is a relatively recent phenomenon. Many young people take hallucinogens as a means of defying parental authority, perhaps because there are no longer any initiation rites at puberty that truly serve as a demarcation between childhood, which is characterized by dependence, and adulthood, which is characterized by independence and the acceptance of responsibility. Other people use drugs to escape from reality, perhaps to forget temporarily the drudgery of a monotonous job or to experience an altered state of consciousness. In modern society, the rituals surrounding the use of many of these drugs have not evolved over a long period of time, and thus the taboos against their use are not elaborate or integral to the successful functioning of the whole culture. In many cases, drug use quickly becomes fashionable, and people find that it is difficult to resist the social pressure of friends and acquaintances who use drugs.

It is often difficult to find accurate, reliable information about the many different types of drugs available today. There is no local "medicine man" who can advise young people. If the drugs in question are also illegal, it is even more difficult to obtain helpful information. The resulting lack of knowledge and abundance of inaccuracies are dangerous, and sometimes deadly. Drug education is sorely needed. Rather than promoting drug use, education can provide an individual with the necessary information to make intelligent decisions regarding drugs and their potential hazards.

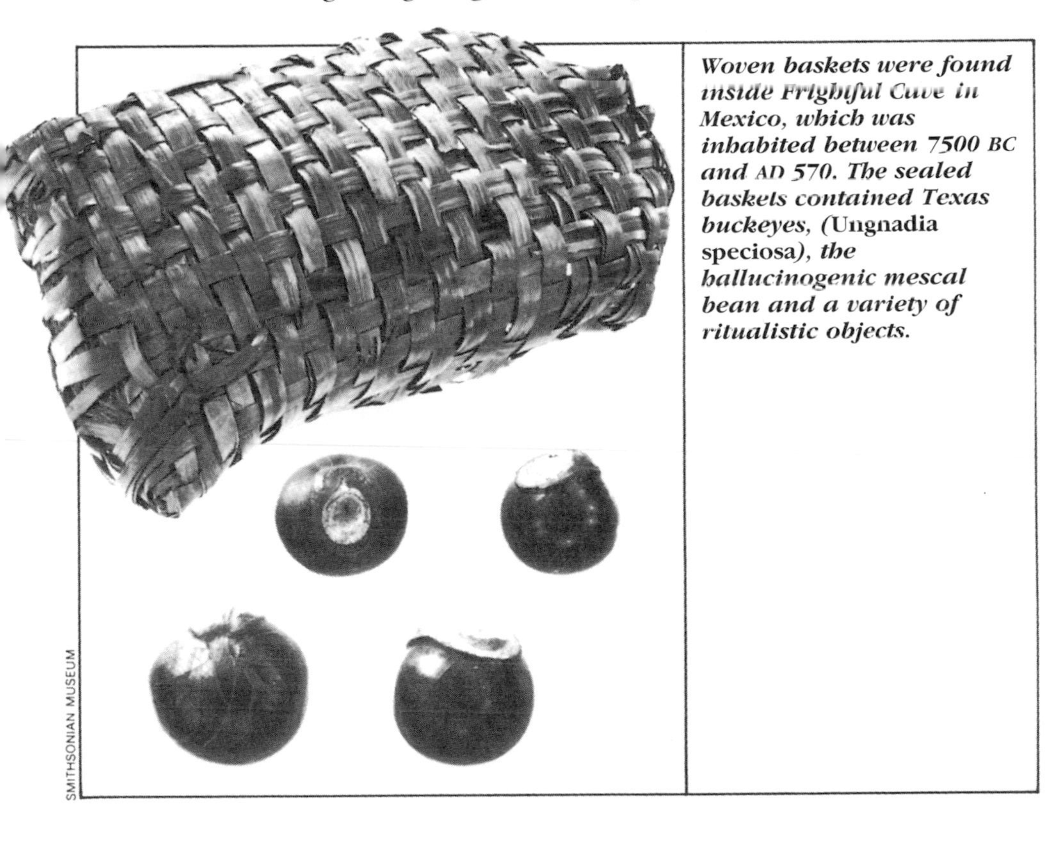


***Woven baskets were found inside Frightful Cave in Mexico, which was inhabited between 7500 BC and AD 570. The sealed baskets contained Texas buckeyes, (*Ungnadia speciosa*), the hallucinogenic mescal bean and a variety of ritualistic objects.***

**Brugmansia suaveolens,** *or tree datura, is cultivated by various South American Indian tribes. The plant contains psychoactive compounds and is used for its medicinal and hallucinogenic properties.*

CHAPTER 1

# FLOWERING PLANTS: HALLUCINOGENS AND THEIR USE

Psychoactive substances are chemical compounds that affect the central nervous system and produce changes in mood, behaviour, and/or intellectual functioning. These substances can be divided into groups of compounds that share similar properties. One such group, *hallucinogenic substances,* or *hallucinogens,* acts upon the brain to produce a variety of abnormal states of consciousness, frequently including colourful visions and/or distorted hearing.

Because all living things are composed of molecules that contain carbon, molecules with one or more atoms of carbon are usually considered *organic* (derived from living organisms). All the hallucinogens found in plants are organic. The chemical structures of some hallucinogenic substances are shown in Figure 1.

## *Alkaloids*

Many hallucinogens are members of a group of chemical compounds called *alkaloids.* Alkaloids are organic compounds that always contain nitrogen and generally cause physiological effects in humans. The type and severity of the effect depends on the specific substance and the dosage that is consumed. Many alkaloids are toxic (harmful to the body), some are lethal, and a few are hallucinogenic. A small amount of an alkaloid may cause hallucinations or sickness, and a large dose may cause death.

Some alkaloids are *teratogenic,* or will produce birth defects or deformities in foetuses, if the mother consumes them during the early stages of pregnancy. These effects may occur even before a woman knows she is pregnant. The people of many cultures have long been aware of the risks of consuming these compounds and have developed ceremonial rules to restrict their use. Because of the danger to the unborn child, the hallucinogenic plants used in various religious ceremonies are generally consumed only by males.

## *The Indole Ring*

Many of the hallucinogens that contain nitrogen are characterized by a common feature called the *indole ring.* This is a chemical structure consisting of eight carbon atoms and one nitrogen atom combined together to form two adjacent rings (see Figure 2). It is no coincidence that the chemical compounds that contain the indole ring are hallucinogenic.

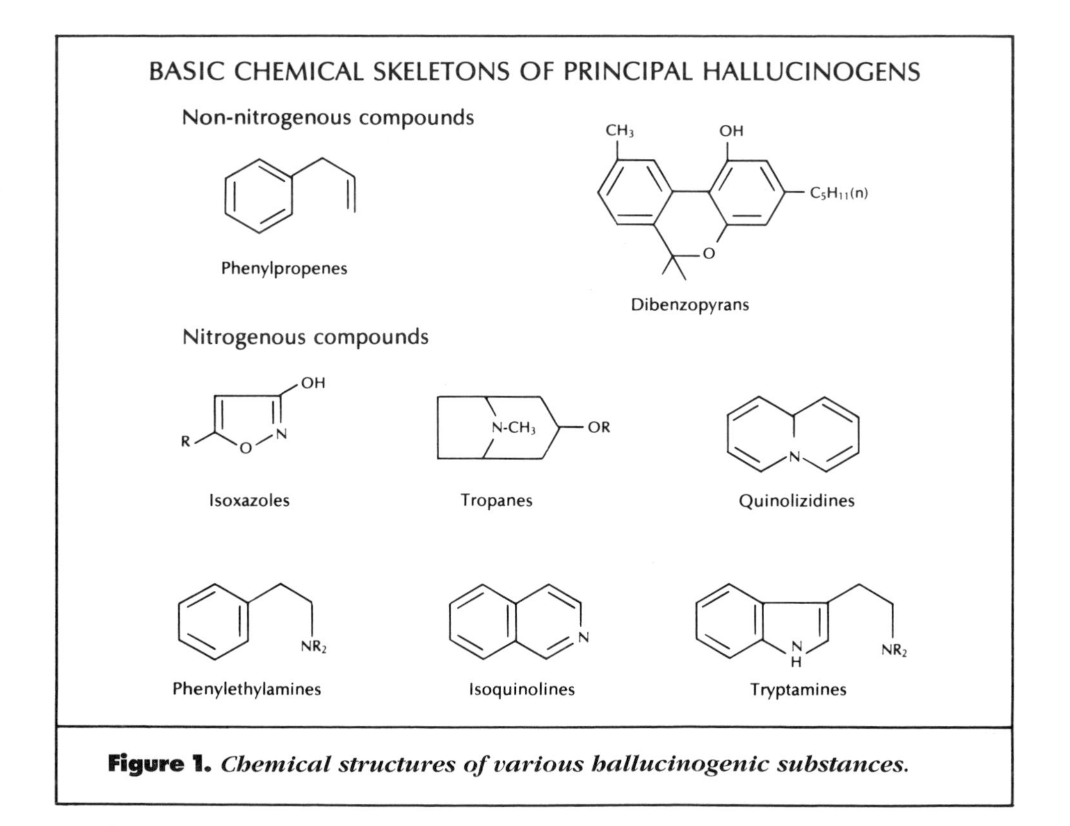


**Figure 1.** ***Chemical structures of various hallucinogenic substances.***

The indole ring occurs in *serotonin,* which is found naturally in the central nervous system of all mammals, including humans, and is involved in the normal functioning of the brain. It is highly likely that hallucinogens containing the indole ring interfere with the action of serotonin, which may explain why these psychoactive substances produce hallucinations.

## *The Classification of Flowering Plants*

Flowering plants belong to the class of plants called *Angiospermae,* characterized by having flowers, which are often showy, and seeds that are enclosed in an ovary that ripens into a fruit. Within this class there are approximately 350 families, only a few of which contain hallucinogenic plants. Each family has a scientific name and a common name. For example, Cannabaceae is the scientific name for the cannabis family. Myristicaceae is the nutmeg family, and Solanaceae is the deadly-nightshade, or potato, family. At the bottom of the hierarchy of plants is the smallest division, the species. There are between 250,000 and 350,000 species of flowering plants. Each species contains only those plants that share certain characteristics and interbreed or have the potential to interbreed. Each species has a scientific name, called a *binomial,* which consists of two words, both of which are written in italics. The first word refers to the genus (a division of organisms that contains closely related species) and the other indicates the actual species. For example, in the binomial *Cannabis sativa, Cannabis* refers to a group of

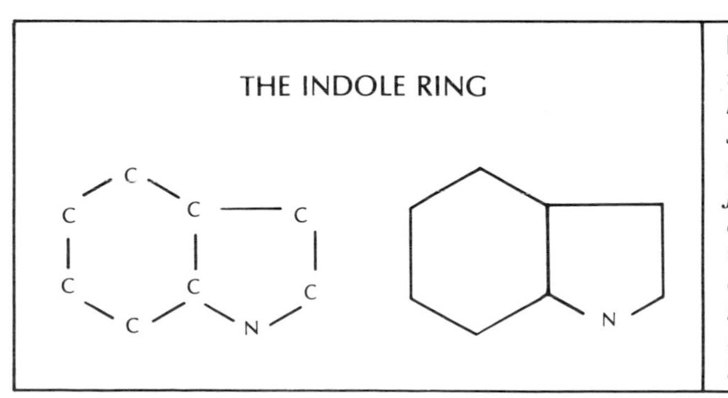


**Figure 2.** ***The indole ring is found in many hallucinogens. Serotonin, which is involved in the normal functioning of the brain, also contains the indole ring, a fact that may explain why the hallucinogens affect the brain and produce hallucinations.***

closely related plants in the genus *Cannabis,* and *sativa* indicates a specific group of marijuana plants within the genus. Similarly, *Myristica fragrans* is the nutmeg species in the genus *Myristica.*

## *Methods of Use*

Hallucinogens occur in various parts of a plant—in exuded matter (such as resin), fleshy fruits, stems, leaves, seeds and roots. The great diversity of both the hallucinogenic compounds and these plant sources has resulted in the development of numerous and sometimes bizarre methods for introducing the active substances into the body. The more common methods include eating, drinking and smoking. The more bizarre techniques include the use of ointments and enemas (solutions introduced into the body via the rectum and the colon). In all cases, the hallucinogenic substances must enter the bloodstream to produce their effects. This process, called *absorption,* occurs most rapidly in areas of the body that are rich in blood vessels, such as the skin, lungs, intestines, mouth, genitals, and rectum.

The hallucinogenic plant may be consumed when it is fresh or after it has been dried. Sometimes the dried plant is ground into a powder and inhaled as a snuff. It may be taken alone or mixed with another plant to increase or alter its effects. If the hallucinogenic compound occurs in a bound form (attached to another chemical), it may be necessary to mix it with lime or ash to create an alkaline (base) solution and free the active substance. If the plant is to be sucked, lime can be placed in the mouth so that the mixing process may take place there. Sometimes the hallucinogenic plant is added to food before or after the food is cooked.

Ointments are usually made by mixing the hallucinogenic plant substance with oils or fats, which increases the rate at which it is absorbed into the body. Although the ointments can be rubbed onto the skin, a faster effect is obtained by applying them to the genitals or anus, where the concentration of blood vessels is greater than in other areas of the body.

Teas, which are made by infusing (steeping) or decocting (boiling) the plant water, can be drunk or taken as an enema. Using an enema allows the absorption of unpleasant substances that if taken orally could cause nausea and/or vomiting. An alcohol infusion, which can increase the concentration of the hallucinogen, can be produced by fermenting an extract of the plant or by adding the plant to wine or beer.

The practice of smoking plant material in pipes, cigars, and cigarettes to achieve psychoactive effects originated in the New World (the home of tobacco), but today this method is used around the world. Each culture has developed its own equipment for doing this, and therefore there is a great variety of smoking devices. Simple pipes may be made from wood, clay, gourds, or animal horns. Water pipes, which originated in the Near East and the Far East and give a cooler and pleasanter smoke, may be constructed out of brass, copper, silver, plastic, or glass. Cigars and cigarettes vary in size from the huge constructions used by inhabitants of the Antilles (a chain of islands in the West Indies) to the thin marijuana cigarettes prevalent in Western cities.

***Swedish botanist Carolus Linnaeus (1707–78), who created the binomial system of classification that is still used to identify all living organisms. Within this system, each organism has a binomial, or a scientific name that consists of two words—a genus name and a species name.***

DR RICHARD E. SCHULTES

**Virola theiodora,** ***a slender tree that grows in Central America and southern Brazil. The bark of this tree contains DMT, a psychoactive drug, and can be heated to produce a hallucinogenic snuff.***

## CHAPTER 2

# THE NUTMEG FAMILY

Within the nutmeg family, or Myristicaceae, there are two genera (plural of genus) that have hallucinogenic species. The Old World genus *Myristica* contains one psychoactive species, the nutmeg plant, *Nyristica fragrans.* The New World genus *Virola,* which grows in the tropics (especially in Amazonia, the region around the Amazon River in South America), contains five important hallucinogenic species—*V. calophylla, V. calophylloidea, V. theiodora, V. elongata,* and *V. peruviana.*

### *Nutmeg*

*Myristica fragrans* is a large evergreen tree that occurs naturally in the Spice Islands (the Moluccas, a group of Islands in Indonesia). Today it is also grown on extensive plantations on these islands as well as on the West Indian island of Grenada. There is no indication that the native people of these islands ever used the plant as a hallucinogen, though there is evidence that it was used for this purpose in India.

Nutmeg trees begin to produce fruit when they are 8 years old and continue to do so for over 50 years. The fruit of the nutmeg looks like a peach and contains a kernel consisting of two parts: the hard seed, commonly called nutmeg, and its fleshy red covering (the aril), which is used to produce mace. Both nutmeg and mace contain three hallucinogenic substances—myristicin, elemicin, and safrole—that belong to a group of chemicals called the *aromatic ethers.* The three substances are related to the synthetic, or laboratory-made, hallucinogens MMDA, TMA, and MDA (respectively), which have been seen on the black market. Scientists believe that safrole is carcinogenic (causes cancer) in humans. The small amounts of nutmeg and mace used in food preparation are probably harmless, but, because of the potentially fatal effects of safrole, the consumption of quantities equivalent to or greater than that in a whole nutmeg (approximately one tablespoon) is not recommended.

When access to other mind-altering drugs has been severely limited, people have resorted to nutmeg. Prisoners, including the American religious leader Malcolm X (imprisoned for robbery between 1946 and 1952), have reported their drug-induced experiences after ingesting as much as a matchboxful of nutmeg.

A drug user must consume large amounts of nutmeg to experience its psychoactive effects. Generally, the hard seed is ground to a fine powder and is either eaten or taken in through the nose as a snuff. Sometimes the powder is heated and the fumes are inhaled, or it is mixed with tobacco and smoked. The consumption of nutmeg causes unpleasant symptoms, such as red eyes, nausea, headache, agitation, hyperactivity, sleeplessness, dry mouth and throat, and motor and speech difficulties. After several hours the drug user may experience euphoria and visual hallucinations. However, it may take several days for the feeling of nausea to pass. During this period the user sometimes suffers from depression, tiredness, headache, and aching bones and muscles.

## *Virola*

The genus *Virola* contains about 60 species, which grow in an area from Central America to southern Brazil. Perhaps 12 of these species are known to contain hallucinogenic substances, though they are only used to produce psychoactive effects in western Amazonia and in adjacent parts of the Orinoco basin. Early in the 20th century, German anthropologist Koch-Grünberg discovered that Indians were using *Virola,* but it was not until 50 years later that its preparation and use was fully described by Richard Schultes.

Most species of *Virola* are woody, and the hallucinogenic species produce a resin-like liquid under their bark that contains hallucinogenic compounds. Though it is sometimes consumed in its raw state, the resin is more commonly subjected to an elaborate process of extraction, drying, and powdering. Occasionally other plant species are added to the powder, which may be consumed as a snuff or shaped into small pellets and taken orally.

Within the genus *Virola, V. theiodora* is the most common source of hallucinogens. It is a slender tree of medium size with smooth brown bark with grey patches. However, the other four *Virola* species mentioned before are also sometimes used by Indians for hallucinogenic purposes.

Initially, researchers assumed that these plants contained the same aromatic ethers that occur in nutmeg. However, further studies showed that *Virola* contains a completely different group of hallucinogens—chemical compounds with an indole ring. These are similarly structured *tryptamines* and *beta-carbolines.* Interestingly, these same compounds occur in *Anadenanthera* (see Chapter 7). The

***A Waiká Indian from Brazil paints darts with a resin made from* Virola, *whose psychoactive drugs can function as effective poisons.***

main components are N,N-dimethyltryptamine (DMT) and 5-methoxy N,N-dimethyltryptamine, both of which are very similar to serotonin, a tryptamine present in the mammalian nervous system. DMT is a very quick acting hallucinogen that causes colourful visual effects within minutes of being sniffed. A deep and disturbed sleep follows the hallucinogenic experience.

Numerous tribes of Indians use snuffs containing *Virola:* western groups such as the Bora of Peru and the Barasana, Kabuyarí, Kuripako, Makuna, and Puinave in Colombia; and eastern groups such as the Karauetarí, Karimé, Kirishana, Pakidaí, Parahurí, Shirianí, Surará, and Yanomami in Venezuela and Brazil. Among the western groups of Indians, the snuff is usually taken by shamans when it is necessary to cure a sickness or when they want to foretell the future. Among the eastern groups of Indians, the snuff is consumed on a regular, even daily, basis. It is used by all males over the age of about 13. Among these people the snuff is not confined to ceremonial use, and excessive amounts are sometimes used.

Ghillean Prance, an ethnobotanist who has studied the Amazon region for many years, observed the use of a *Virola*

***A Puinave man cooking the* Virola *resin, from which he will produce yato, a hallucinogenic snuff that contains DMT. When inhaled, the drug produces effects that come on very suddenly and can be quite overwhelming.***

snuff by the Sanama Indians of Brazil. The Indians select a wild *V. theiodora* and strip off the bark, which they heat over a fire until a resin oozes out. The resin is then dried and powdered to form a snuff. Among these people, the snuff is used by the shaman when he wishes to treat a patient, or it is used in a ceremony after the death of a tribe member.

The death ceremony, which may last up to eight days, commences with a hunt for game such as tapir and wild pigs. After the hunt, there are several nights of dancing. During one of these nights the hallucinogenic snuff is used, either by taking a pinch and sniffing it or by having it blown into the nostrils by a small blowpipe. All the participants then begin to dance and shout. This is followed by a vigorous chest-hitting ceremony, during which grievances are settled. The Indians sometimes use rocks or pointed metal to strike their blows, and blood is often spilled, but the snuff appears to act as an effective anaesthetic (painkiller). After the cermony, the participants squat and shout deafeningly into their neighbour's ear. At the peak of excitement, the ashes of the dead person are poured onto a fire. The shouting continues until the effects of the *Virola* snuff diminishes.

***The preparation of hallucinogenic plants usually involves elaborate rituals. Here a Waiká man crushes dried*** **Virola** ***resin.***

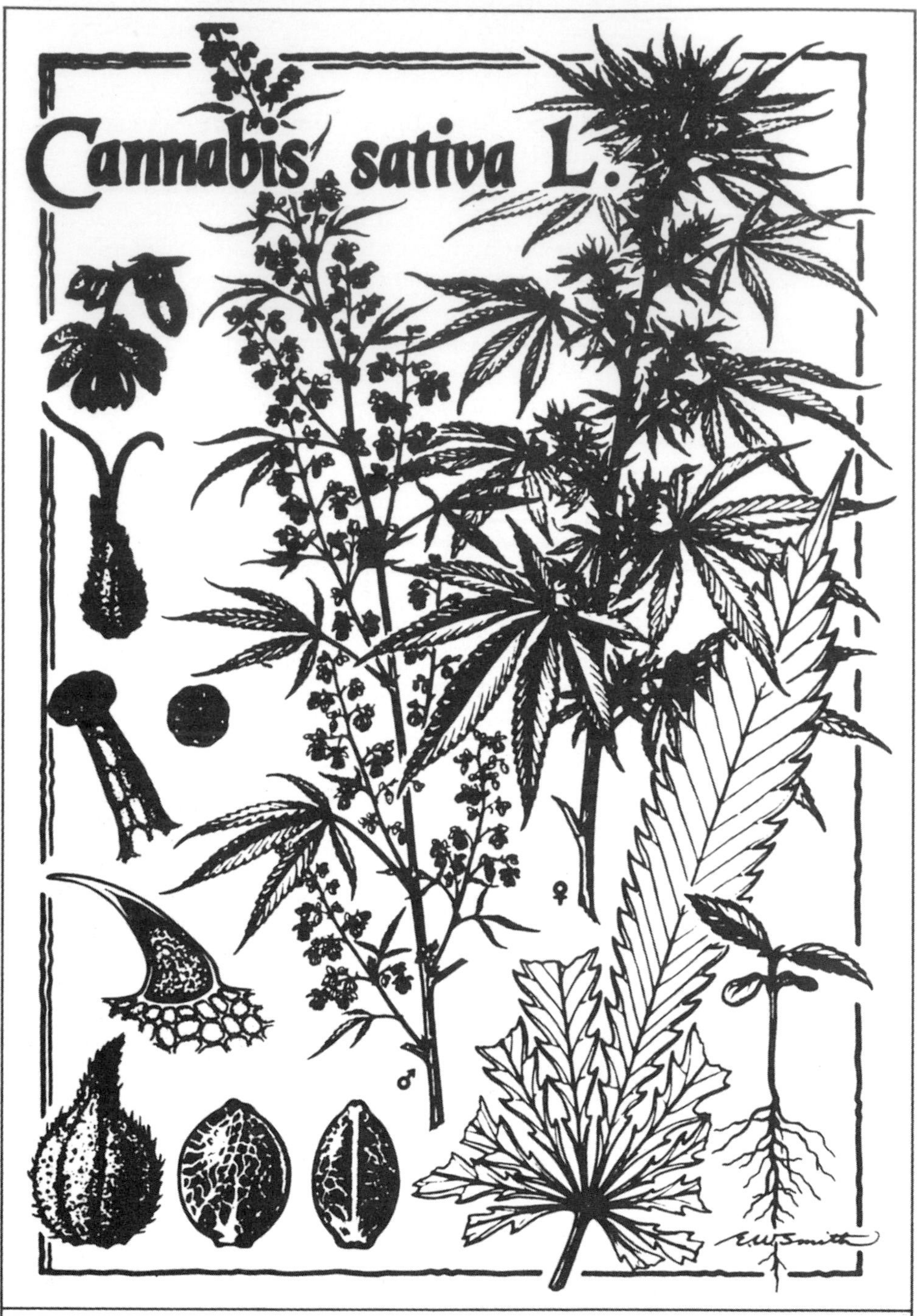


DR. RICHARD E. SCHULTES

***A drawing depicts the various properties of* Cannabis sativa*, commonly known as hemp or marijuana. Products made from this plant have been used and abused by human culture for at least 8,000 years.***

# CHAPTER 3

## THE CANNABIS FAMILY

The cannabis famiy, or Cannabaceae, contains two genera, *Humulus* and *Cannabis.* Both originated in the Old World and have been used by people for thousands of years. The flowers of *Humulus lupulus,* or the hop plant, have a resin that is used as a preservative and a natural flavour in beers made from malted barley. The active ingredients in the resin are the bitter-tasting compounds called *sesquiterpenes,* which inhibit the growth of bacteria.

Members of the genus *Cannabis* are known by many names, including cannabis, hashish, hemp, and marijuana. The cannabis plant is a source of many useful products: strong fibre for ropes, cloth, and paper; nutritious fruit; industrial oil and medicine. Though scientists know that this plant has been associated with human culture for at least 8,000 years, the exact details of its history, including when and where it was first grown by farmers, are not known.

Cannabis is a *dioecious* plant. This means that individual plants are either male or female. Both sexes of the plant are used for fibre and resin production, though female plants are generally considered to be the better producers of resin, partly because male plants dry up soon after flowering. When the male and female plants are separated before pollination (the transfer of pollen and subsequent fertilization of the plant), the female plants flower but do not produce seeds. Such *sinsemilla* plants (Spanish for "without seeds") are said to produce the greatest quantity of resin. The sticky resin occurs in glandular hairs, which are most abundant on the flowering heads and surrounding leaves. The resin contains the hallucinogenic compounds called *cannabinoids.* The major active compound is delta-1-tetrahydrocannabinol, also called delta-1-THC or just THC.

Over thousands of years, different strains of cannabis have been selected as sources of fibre and resin. The fibre usually comes from plants called hemp, and the resin from plants often called marijuana. Fibre plants tend to be tall and have a low content of the hallucinogenic cannabinoids. Resin plants are generally shorter and have a higher content of cannabinoids. The process of selection has led to confusion about the correct scientific name for cannabis. Some botanists believe that *cannabis sativa* is the name for the fibre plant (hemp) and *Cannabis indica* is the name for the resin

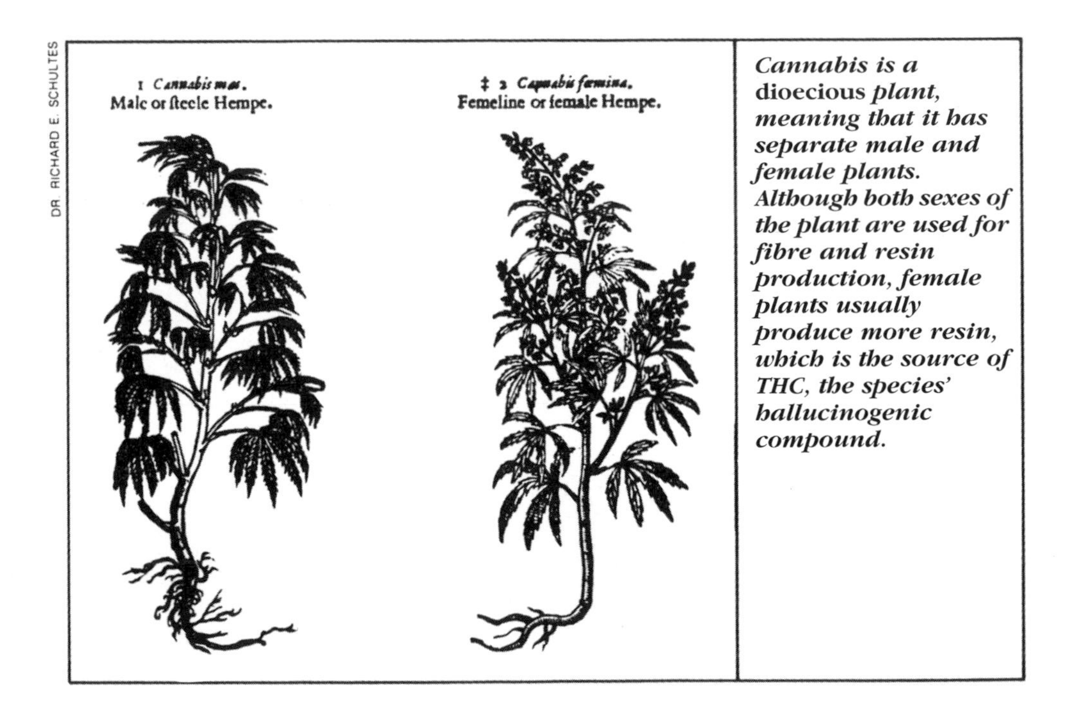


***Cannabis* is a *dioecious* plant, meaning that it has separate male and female plants. Although both sexes of the plant are used for fibre and resin production, female plants usually produce more resin, which is the source of THC, the species' hallucinogenic compound.**

DR RICHARD E. SCHULTES

plant (marijuana). Other botanists believe that the name *Cannabis sativa* refers to both kinds of plants. However, recent examinations of the internal anatomy of cannabis have provided unquestionable evidence that there are at least two separate species.

This may seem to be a purely botanical argument, but it has important legal implications. In many parts of the world the possession and use of cannabis is illegal. However, most laws specifically prohibit *C. sativa* and do not even mention *C. indica.* Several defendants in court have successfully argued that they possessed *C. indica* and were therefore not breaking the law. This method of defence will become ineffective when the laws are changed to include all members of the genus *Cannabis.*

### *How Cannabis Is Used*

There are several ways to introduce cannabis and its extracts into the body. The leaves or flowering tips can be dried to

DR. RICHARD E. SCHULTES

**A sinsemilla *plant. Because these plants have a great amount of resin,* sinsemilla *(Spanish for "without seeds") contains large amounts of THC.***

form what is commonly known as *grass,* which can be smoked, drunk as a tea, or mixed with food. The cannabinoids can be obtained in a more concentrated form by collecting the resin from the flowering heads. The resin can be smoked, either alone or mixed with tobacco, or it can be used in cooking. It is also possible to make an alcoholic extract of the resin, which produces a potent solution known as cannabis oil. The oil is mixed with tobacco and smoked.

Of all the hallucinogenic plants, cannabis has the most variable effect on people. While it can produce dreaminess, changes in perception, and a sense of well-being, cannabis can also lead to *psychological dependence* (a condition in which the drug user craves a drug to maintain a sense of well-being and feels discomfort when deprived of it).

### *History of the Use of Cannabis*

The use of cannabis has been traced back to a time as early as 8000 years ago. At this time a culture on the island of Taiwan

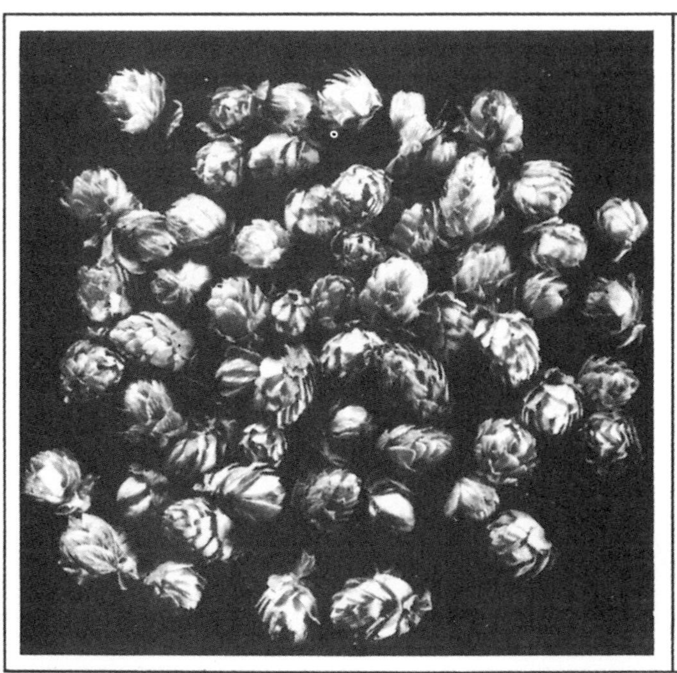

***The hop plant, or* Humulus lupulus, *like marijuana is a member of the cannabis family. It produces flowers that have a resin that is used as a preservative and as a natural flavour in beers made from the malted barley.***

decorated its pots by pressing strips of hemp rope into the wet clay. Some archaeologists believe that the Chinese were the first to use cannabis for its medicinal effects, around 4000 BC. If the cannabis plant originated in Central Asia, as scientists who have studied the plant currently believe, then it must have been transported to China at an even earlier date. Therefore, its use elsewhere must necessarily precede its earliest recorded use in China. The fibre markings on pots are indirect evidence of the use of cannabis. The oldest preserved specimens of the plant itself are pieces of cloth found in a Chinese burial site dated approximately 1200 BC. It is thought that in early China the wealthy people wore clothes made from silk, while the poorer people wore cloth woven from hemp fibres.

The Chinese symbol for cannabis originally meant "tall plant", though the symbol gradually changed to mean "tall plant growing near a house". There are specific Chinese symbols to indicate male and female cannabis plants as well as their flowers, fruits and seeds. Cannabis fibre (hemp) provided stronger bowstrings than bamboo fibres, which were formerly used. The Chinese were also the first to use cannabis fibres to produce paper, which replaced heavy wooden tablets and expensive silk fabric. A mixture of cannabis fibres and mulberry bark was crushed in water, and when the mixed fibres floated to the surface they were dried in moulds to produce sheets of paper. This process dates from the 1st century BC. It was a thousand years before the knowledge of papermaking reached the Western world via Arabia and the Moorish Empire.

The first written account of the use of cannabis in Chinese medicine is thought to come from the 1st century. The female plants were preferred for medicinal use due to their being predominantly *yin.* (According to the Chinese system, everything has both *yin* qualities, characterized as passive, negative, and feminine, and *yang* qualities, characterized as active, positive, and masculine.) Thus, to treat ailments such as rheumatism (pain in the joints, and inflammation, soreness, and stiffness of the muscles), constipation, and absentmindedness, which were attributed to a loss of *yin* from the body, an extract from a female plant was used. By the 2nd century cannabis extracts were used as an anaesthetic for surgical operations involving the stomach and intestines, procedures for which acupuncture was ineffective. (Acupuncture is a Chinese method of decreasing or eliminating pain by inserting fine needles into specific points of the body.)

It is not known when the Chinese first used cannabis to achieve its hallucinogenic effects, but it is known that consuming the drug for this purpose was disapproved of in the period around 600 BC. However, in the 1st century the seeds of the plant were still being burned in incense burners to give people the ability to see spirits and achieve immortality. Cannabis was used in this way by only a small segment of the population, and thus it never became a social problem, such as the one later caused by the use of opium.

Cannabis also has a long history of use in India, where it is often known as *bhang* (a drink made from the leaves),

DR. RICHARD E. SCHULTES

***Before the Civil War, cannabis was widely cultivated in the southern United States for its fibre, which was used to make rope. Although rope production declined after the war, the plant spread like a weed.***

*ganja* (the dried, resin-rich flowering heads), or *charas* (the pure resin). An Indian medical book written before 1400 BC described the value of cannabis in the treatment of anxiety. There are numerous references to the pleasurable use of cannabis in the literature of India, where use of the plant for its hallucinogenic properties became, and indeed still is, well established in some Himalayan regions. The use of cannabis was often associated with the goddess Kali, and *bhang* was sometimes ritually consumed before sexual intercourse. In the early 20th century the British authorities concluded that the use of cannabis in India was such an integral part of the people's life that it could not and should not be prohibited.

An ancient Greek manuscript describes how the Scythians, a group of nomadic tribes that lived in an area stretching from the Danube River region of Europe to Siberia, used cannabis around 600 BC. Archaeological evidence had confirmed that at funerals these people dropped cannabis seeds on red-hot stones and breathed the vapours that were released. In addition, sometimes people were buried with

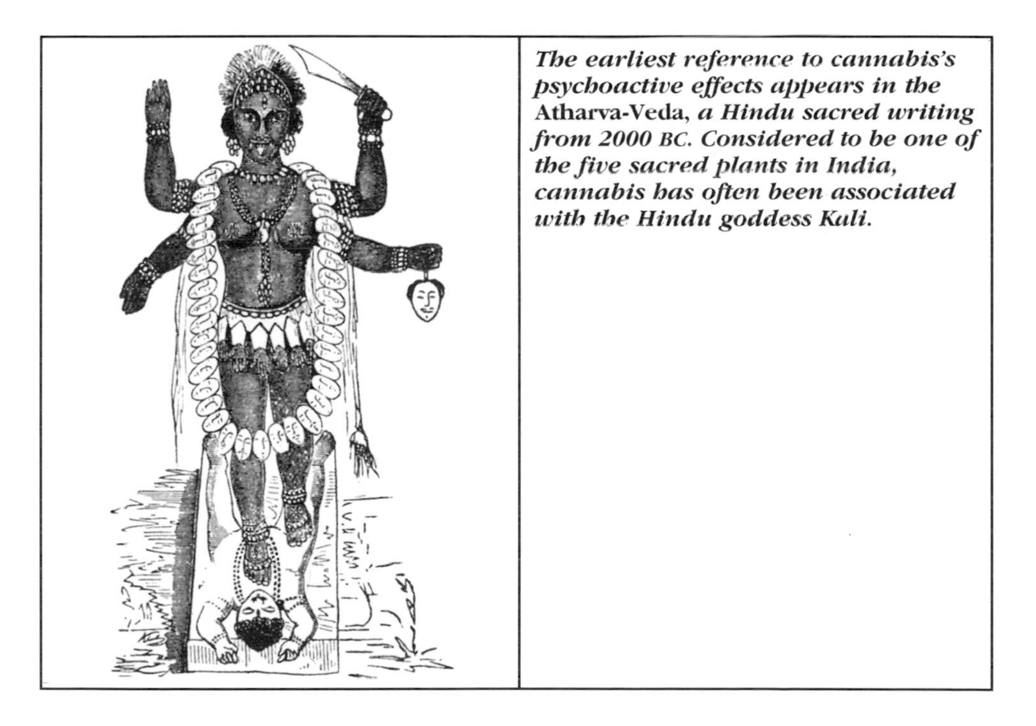

***The earliest reference to cannabis's psychoactive effects appears in the*** **Atharva-Veda,** ***a Hindu sacred writing from 2000 BC. Considered to be one of the five sacred plants in India, cannabis has often been associated with the Hindu goddess Kali.***

their own smoking outfit, which consisted of a bronze cauldron filled with stones and hemp seeds and a leather canopy, or tentlet. The descendants of the Scythians who live in Eastern and Western Europe still honour the dead by throwing cannabis seeds onto fires during harvest times—a tradition unbroken for over 2,500 years.

The spread of cannabis to the West was slow and uncertain. Cannabis fibres have been found in sites in Turkey dating from 1000 BC. The first reliable Egyptian reference to cannabis is in the 3rd century. The ancient Greeks grew cannabis for its fibres as early as 600 BC, but their literature makes few references to the use of the plant as a medicine or a hallucinogen. An ancient Roman herbal (a book describing medicinal plants and their uses) from AD 70 mentions the use of cannabis seeds to treat earache, and another book from a hundred years later mentions a pleasurable dessert made from cannabis seeds. Apparently, cannabis was not cultivated in Rome, and little use was made of its hallucinogenic properties, but the fibres were much used for ropes for the navy. It was as a fibre plant that cannabis was introduced to the rest of Europe at a later date.

Perhaps the most remarkable story about the use of cannabis was told by Marco Polo, the famous 13th-century Italian traveller and merchant who went overland to China. In northern Iran he had learned of a band of thugs known as the Assassins, whose leader supposedly used cannabis to control his followers. The members of this murderous group were rewarded with *hashish* (resin-covered cannabis flowers) and were known as *hashishins*—later modified to *assassins,* a term that now means a type of murderer. Marco Polo's story had never been confirmed, but it has often been employed by uninformed people to indicate that cannabis turns people into murderers. Cannabis does have a long history of use in the Arab world, but it is most frequently consumed to escape the harshness of life, rather than to gain inspiration for violence.

The cannabis plant has spread throughout the world. Even where cultivation has ceased, it often continues to grow as a weed. Evidence from chemical examination of the residues in pipes shows that people smoked cannabis in the 13th and 14th centuries. Arab traders introduced cannabis to the people along the east of Africa, reaching southern Africa in pre-Portuguese times (before 1500). Its use then spread west and south from the Zanzibar region by way of various herders and traders. Cannabis became common in many parts of southern Africa and has been consumed for its hallucinogenic properties by the people of many different cultures. In fact, colonists often paid their native labourers with cannabis, though this practice has now largely disappeared. The plant reached Southeast Asia in the 16th century, and in Kampuchea (Cambodia) it is still employed as a hallucinogen and a medicine.

It is not known for certain how cannabis spread to Europe, though there is a clear link to the history of the Scythians. There are many Eastern European customs that involve the use of cannabis. In Poland and Lithuania a soup was made from cannabis seeds on Christmas Eve, when the dead are believed to visit their families. In Latvia and the Ukraine a cannabis dish was prepared on Three Kings Day. The height of the cannabis plants was of great importance in

THE BETTMANN ARCHIVE

*A portrait of the Italian traveller Marco Polo (1254–1324) by the 16th-century Venetian artist Titian. In his journeys through northern Iran, Marco Polo learned of a band of thugs whose leader used cannabis to control his followers. The members were rewarded with* **hashish** *(resin-covered cannabis flowers) and were thus known as* **hashishins**, *a term later modified to* **assassins.**

fibre production, and the European peasants, hoping that the plants would grow tall and yield long fibres, planted the seeds on the days of the saints who were believed to have been tall. In Germany cannabis seeds were thrown into the air as the plant was sown, and in some areas the planting was done around "high" noon. In Poland married women leaped high into the air during the Hemp Dance, performed on Shrove Tuesday. The southern Slavs believed that cannabis at weddings would bring wealth and happiness. Therefore, not only was the new bride showered with cannabis seeds, but she stroked the walls of her new house with cannabis fibres.

Initiation ceremonies may often accompany the harvesting of cannabis, and there is often much festivity. In Switzerland young men wear fancy masks and tempt young ladies with offers of gifts. In the Ukraine, it was thought that hemp gathered on Saint John's Eve had the magical property of deterring evil actions.

Cannabis also has a place in the folk medicine of Europe.

THE BETTMANN ARCHIVE

***An early 18th-century wood engraving entitled "A Female Tobacco Manufacturer". The woman, a tobacco flower in one hand and a snuff box in the other, is dressed in a Turkish tobacco-leaf hat, blouse and skirt. Although the cultivation of tobacco and the production of tobacco-containing products have great economic importance, the use of these products is extremely hazardous.***

In Germany, cannabis sprigs were laid on the body to treat convulsions or aid a difficult childbirth. In Poland, Russia, and Lithuania, the vapours released from seeds dropped onto hot stones were used to treat toothaches. In the Ukraine, people ate hemp porridge for the same purpose. Cannabis was given as a treatment for fevers in Czechoslovakia and Poland, where it was also used to dress wounds. In Russia, cannabis-seed oil was used for rheumatism and jaundice. Serbians believed cannabis to be an aphrodisiac.

Cannabis was not generally cultivated in England due to opposition from farmers, who claimed that the plant exhausted the soil and produced a foul smell when it was retted (placed in water to decompose and release the fibres). Still, the English navy had a huge demand for hemp ropes, and it tried to fill this need by ordering the colonists in Virginia and Maryland to grow cannabis as a fibre plant in early 17th century. The colonists preferred to grow tobacco because it

UPI/BETTMANN NEWSPHOTOS

***Workers grinding up tobacco leaves for smoking, chewing and snuff. Tobacco was introduced into Europe in the 16th century by Spanish explorers returning from the New World.***

was more profitable, and therefore very little hemp fibre was sent to England.

The Pilgrims grew cannabis for fibre, which was used to produce clothing and rope. Ropewalks were constructed in Boston, and ropemaking helped the local shipbuilding industry. During the American Revolution the colonist could no longer import clothing from Europe, and thus cannabis fibre became essential. The hemp-fibre industry, which eventually became centred in the South, required a great many workers and so increased the demand for slaves. After the Civil War, hemp production declined as labour costs rose, though the cannabis plant continued to be cultivated as a source of birdseed, especially in the Midwest. The cannabis plant quickly spread beyond the cultivated areas and now grows as a weed. In 1969 there were an estimated 150,000 acres of the wild plant in Nebraska alone.

Cannabis's medicinal properties were also recognized in the United States, where they were first mentioned in an

THE MARIHUANA PROBLEM

IN THE CITY OF NEW YORK

Sociological, Medical, Psychological
and Pharmacological Studies

by the

MAYOR'S COMMITTEE ON MARIHUANA

THE JAQUES CATTELL PRESS
LANCASTER, PENNSYLVANIA

DR. RICHARD E. SCHULTES

*The title page from* **The Marijuana Problem in the City of New York,** *published by the Mayor's Committee on Marijuana in the late 1930s. The report stated that marijuana did not lead to the use of other drugs, and that the sociological, psychological and medical dangers commonly linked to marijuana had been exaggerated. However, more recent research has discovered much evidence to the contrary.*

American medical book in 1843. Eventually cannabis was officially recommended as a treatment for a number of ailments, including gout, tetanus, hysteria, depression, and insanity. Cannabis extracts were prescribed for dysentry (an intestinal disorder), which was prevalent among Civil War soldiers. And until 1937 cannabis was listed in the *US Pharmacopoeia* as a tranquillizer. Users were only warned against consuming large doses. Psychiatrists also experimented with treating their patients with cannabis.

However, cannabis had several disadvantages: its concentration in medicinal preparations varied from one supplier to the next; it was not soluble (did not dissolve) in water and therefore could not be injected to produce its effects rapidly; and the effects varied from one patient to another. Cannabis never found much favour with doctors, but it was widely available in patent medicines (packaged, nonprescription drugs with secret ingredients) until government approval was removed in 1941.

### *Cannabis in the 20th Century*

In the early part of the 20th century an estimated 1 million Americans were addicted to either opium or morphine, two narcotics (drugs that depress the central nervous system, especially those derived from the opium poppy) that were easily available through the mail, from travelling medicine men, and from ignorant doctors. At this time cannabis extracts were being sold as cures for addiction to these narcotics. Eventually the authorities acknowledged the severity of the opium problem in the United States. Various state laws restricting narcotics were passed, but because exemptions were made from patent medicines, the problem continued. Chinese Americans, chosen as scapegoats, were unfairly blamed for the opium problem. This practice set a precedent, and during later antidrug campaigns other minorities were often singled out.

In 1970 the Comprehensive Drug Abuse Prevention and Control Act lowered the penalties for cannabis possession. Presidents John F. Kennedy, Lyndon Johnson, and Richard Nixon appointed commissions to investigate cannabis, and all of them came to the conclusion that the drug-containing plant was much less harmful than people had previously believed. Since then, several states have decriminalized the use of cannabis. Other states have made it legal to smoke cannabis for the treatment of glaucoma, an eye disease characterized by an increase in pressure within the eye that causes a gradual loss of sight. The cannabinoids in the plant lower the pressure in the eyeball and thus prevent blindness.

While some people argue that cannabis is no more harmful than alcohol and tobacco, in some ways cannabis appears to combine some of the worst aspects of both of these drugs. Chronic smoking of marijuana can cause lung cancer, and produce hallucinations, impair cognitive abilities, and distort the user's sense of time. This can, and often does, lead to the same types of accidents that result from the consumption of alcohol. Consequently, at the present time, cannabis is regarded as an illegal drug in most countries. The

***Because they have a highly developed sense of smell, dogs can be trained to detect hashish, which has a distinctive, strong, sweet smell. This is just one of the many methods used by drug enforcement officials to curb the smuggling of illegal substances.***

Single Convention on Narcotic Drugs was adopted in 1961 and 115 countries became signatories. The Convention aims to ensure that controlled substances, including cannabis, are used exclusively for medical and scientific purposes. Cannabis use, production and trafficking are prohibited.

In the United Kingdom the Misuse of Drugs Act, 1971, places cannabis in class B, with heroin and cocaine in class A. Maximum sentences are 14 years' prison and a fine for trafficking and 5 years' prison and a fine for possession. In practice, most offences relate to possession of small amounts for personal use and are dealt with the Magistrates' Courts by small fines. See *Marijuana: Its Effects on Mind and Body* in this series.

***Aircraft are regularly searched by members of staff, before and after flights, to ensure that there are no smuggled packages on board.***

DR. RICHARD E. SCHULTES

*The San Pedro cactus,* Trichocereus pachanoi, *contains the hallucinogen mescaline and was used in northern Peru by shamans for divination and healing over 3,000 years ago.*

# CHAPTER 4

# THE CACTUS FAMILY

The cactus family, or Cactaceae, is known to most people as a group of tropical desert plants. But some cactus plants, or cacti, grow as far north as British Columbia, on the peaks of the 10,000-foot Andes, or in the jungle. There are approximately 50 genera of Cactaceae, all of which originated in the New World, and about one-tenth of them contain hallucinogenic species. The most famous hallucinogenic cactus is peyote, which is revered by the members of the Native American Church, a religious group formed by American Indians. Another well-known hallucinogenic cactus is called San Pedro, which formed the basis of magico-religious ceremonies in Andean South America for at least 3,000 years.

Both peyote and San Pedro contain, among other psychoactive substances, hallucinogenic compounds called *phenylethylamines.* These alkaloids also occur in other genera in the cactus family, but only a few of the plants are consumed for their hallucinogenic properties. *Mescaline,* a phenylethylamine, has the most potent hallucinogenic properties and is found in both peyote and San Pedro.

## *Peyote*

Peyote is a member of the genus *Lophophora,* which contains two species. The most widespread species is *L. williamsii,* or true peyote, which is native to northern Mexico and adjacent parts of Texas. The other species, *L. diffusa,* is native to only a small area of northern Mexico.

Though this species is hallucinogenic, it does not contain mescaline.

Peyote does not resemble a typical cactus. The plants are small with tufts of hair, but have no significant spines. The succulent heads can measure up to four inches in diameter. They often occur in clumps of smaller heads, though this is not the natural condition but the result of harvesting practices. When a peyote collector slices off the head, the parsnip-like root left in the ground sprouts several new heads. In this way, the same plant can be harvested many times.

Peyote can be eaten fresh or dried to form *mescal buttons,* which remain psychoactive for many years. The buttons can be moistened and then eaten or, because mescaline is soluble in water, they can be made into a tea. Consumed in either form, peyote produces physical effects, such as nausea and vomiting, and psychological effects, such as visual and auditory hallucinations, feelings of weightlessness, and synesthesia (a mixing of senses; for example, a sound may produce a mental image).

DR. RICHARD E. SCHULTES

***A painting of a Kiowa Indian roadman by Kiowa artist Stephen Mopope. Archaeological evidence indicates that peyote, a cactus that contains the hallucinogen mescaline, was used ceremonially in Texas at least 3,000 years ago.***

The Spaniards encountered peyote when they first came to the New World. They associated the plant with the Aztecs' bloody sacrificial rites and called it "the devil's root". Eventually the Holy Office of the Inquisition enacted the first antidrug laws of the New World, and in 1690 the use of peyote was condemned as being associated with superstitious behaviour, which was contrary to the missionaries' Christian beliefs. In an attempt to suppress the use of peyote in religious ceremonies, the Spaniards tortured and even killed many Indians. They did not understand how difficult it would be to suddenly abolish rituals that had been practised by the natives of some regions for thousands of years. The Spaniards only succeeded in forcing the Indians to be extremely secretive about their ceremonies. Because of this, many of the facts about the ritual use of peyote have only recently been rediscovered. It was only in the 1960s that a Western scientist first observed a traditional peyote pilgrimage by the Huichol Indians.

Although peyote grows only in a limited area, its use has spread elsewhere. Indians as far north as Canada now use peyote in their religious ceremonies. Sometime in the 19th century the Kiowa and Comanche Indians learned about peyote from neighbouring Mexican Indians and adopted the

***An Osage peyote church in Oklahoma. In the 1890s, after travelling into Oklahoma and participating in peyote rituals, anthropologist James Mooney encouraged the formation of a unified Indian movement which would protect the members' right to use peyote. As a result, in 1918 the Native American Church was incorporated.***

peyote cult as a new religion. It rapidly spread to other tribes of Plains Indians. Missionaries and local governments viewed the peyote cult as contrary to Christianity, which they wished to impose on the Indians. Because many anthropologists and other scientists defended the Indians' use of peyote, federal laws prohibiting the cult were not established. In an attempt to escape persecution, the Indians formed the Native American Church, which teaches brotherly love, high moral principles, and abstention from alcohol. In 1985 the church had a quarter of a million members.

Despite the US Constitution's guarantee of separation of church and state, a few states still have laws against the use of peyote in religious ceremonies. In 1888 government agents attempted to ban the use of peyote by Indians, and 15 states later passed laws to endorse this prohibition. Many of these state laws were later repealed because a federal ban on peyote never materialized. However, in recent times, Native American users of peyote have been arrested in Arizona and California. Other religious organizations continue to harass the members of the Native American Church, in spite of the fact that there is no indication that the ceremonial consump-

DR RICHARD E SCHULTES

**Ariocarpus retusus,** ***or false peyote. Although this rectangular-leafed cactus contains hallucinogens similar to mescaline, Indians consider it dangerous because it is extremely potent and "drives people mad".***

tion of peyote leads to addiction or that it is harmful to the user's health.

Peyote rituals vary from tribe to tribe. For example, though most North American Indians get their cacti by mail from Texas, the Mexican Indians have the added ritual of the peyote hunt. The Huichol Indians live in the western Sierra Madre mountains of Mexico, and, since peyote is not native to this region, they go on an annual 600-mile pilgrimage to collect it. Traditionally, the journey was made on foot, though today the *peyoteros,* or peyote hunters, employ modern methods of transportation. A ritual confession and purification precede the pilgrimage, and along the way all the sacred sites must be visited. On arrival at the peyote site, any new peyote hunters are ritually blindfolded, and the shaman tells peyote stories and legends. According to ritual, the first peyote plant to be discovered is shot with arrows. Only after the completion of the complex ceremony are the partici-

PETER FURST

***The Huichol Indians, believing that peyote possesses a soul and should not be bruised, pack the cacti carefully in baskets and woven bags.***

pants allowed to gather the peyote, which they do by cutting off the tops of the plants. Some peyote is eaten immediately, but most of it is dried for later use.

The members of the Native American Church have regular peyote ceremonies as well as special ones for births, birthdays, and deaths. Meetings are often held on Saturday nights and last all night. Various taboos, such as abstaining from salt before and after a meeting, are sometimes observed. Women are usually allowed to consume peyote, but they must refrain from the ritual singing and drumming. A typical Plains peyote ceremony may take place in a tepee, in which there is a crescent-shaped altar and a small fire. The ritual equipment includes feathers, sage, tobacco, a drum, a staff, cedar incense, a fire stick, a gourd rattle, an altar cloth, an iron pot for water and an eagle-bone whistle. There may be as many as 30 participants, who sit in a circle around the central fire and altar. A ritual cigarette precedes the burning of incense, after which the sage and peyote are passed. Each participant consumes four peyote buttons. Drumming and singing commence and continue until midnight, when the whistle is blown in the directions of north, south, east, and west. This is followed by further singing. The participants

***In these Cheyenne peyote ornaments (above), the crescent represents the peyote altar, and the star the peyote plant. The Cheyenne ceremony often takes place in a teepee within which there is a crescent-shaped altar and a small fire. Different types of feathers (left) are among the ritual equipment used during the ceremony.***

may consume more of the peyote at any time. At dawn, the ceremony is formally over and food is consumed.

### *San Pedro*

San Pedro (*Trichocereus pachanoi*), which also contains mescaline, is a large columnar cactus that can grow up to 20 feet tall and has from four to seven longitudinal ribs. It naturally occurs in the Andes mountains of South America, though it is also cultivated by the natives of Bolivia, Ecuador, and Peru. Archeological evidence suggests that San Pedro was used over 3,000 years ago. One stone carving, dated 1300 BC is of a local deity holding the cactus, and ancient cloth from approximately the same period shows the cactus with jaguars and hummingbirds. Pottery from a somewhat later time includes the plant in association with spirals, which are thought to be representations of the drug's mental effects.

The seven-ribbed San Pedro is usually used by the *curandero* (the local name for the folk healer or shaman), whose ceremony combines Christian and ancient pre-Columbian beliefs. The cactus is cut into one-inch thick slices, which are boiled in water for up to seven hours to produce what is called *cimora.* Powdered bones, cemetery dust, or other plants (such as *Datura,* a hallucinogenic member of the nightshade family) may be added to the boiling potion.

Cimora is used in rituals during which people foretell the future. It is also used by the *curandero* to help find the cause of sickness or other troubles. Each *curandero* possesses his own *mesa* (a collection of Christian and satanic artifacts), which are laid out on an altar cloth during the curing session. In northern Peru the session consists of two parts: the ceremonial ritual, which is performed before midnight, and the curing ritual, which follows it. The session takes place in the open on any night except Monday (the day when the spirits of the dead are on the prowl) and includes prayers, invocations, and chants. At midnight the *curandero* drinks a cup of the hallucinogenic *cimora.* Then each patient takes his or her turn in front of the mesa, and the *curandero* "sees" the cause of the patient's trouble. After a final act of purification, the session is over.

DR. RICHARD E. SCHULTES

*A Tukano boy ready to undergo* Yurupari, *or puberty rites. The ceremony includes the drinking of the hallucinogenic drink* caapi, *and displays of courage, which involve painful whippings.*

# CHAPTER 5

# THE BARBADOS CHERRY FAMILY

The Barbados cherry family, or Malpighiaceae, grows in the tropics of the Old World and the New World, especially in South America. Within this family there are at least two genera—*Banisteriopsis* and *Tetrapteris*—that contain hallucinogenic species used by the Indians in the Amazon-Orinoco area of South America. These plants can be cultivated to maintain a constant supply, but the older, wild-growing lianas (vines) are considerably superior to the cultivated plants.

## *Banisteriopsis*

The bark from the two lianas *B. caapi* and *B. inebrians* contains the psychoactive substance. There is unconfirmed evidence that some groups of Indians simply chew the plant material to achieve the hallucinogenic effects and that others use the bark to prepare a hallucinogenic snuff. But most commonly the bark is used to prepare a hallucinogenic drink, which is known by such names as *caapi, kahpi, mihi, dapa, pinde, natema, yajé,* and *ayahuasca,* which means "vine of the souls". The drink is very bitter and causes nausea. It can be prepared either by boiling pieces of the bark in water or by squeezing the bark in cold water. The plant source of *ayahuasca* was first identified in 1851 by Richard Spruce, a famous English explorer.

The hallucinogens present in *ayahuasca* belong to a group of compounds called harmala alkaloids, or beta-carbolines. Harmaline and, to a lesser extent, harmine and d-

1,2,3,4-tetrahydroharmine are the major psychoactive ingredients. The effects produced by drinking *ayahuasca* vary from person to person and depend upon the other plants that may have been added to the drink. It can produce physical effects, such as nausea, vomiting, intestinal purging and increased blood pressure and heart rate, and psychological effects, such as visual and auditory hallucinations. Intoxication may end with deep sleep and dreams, and the following day the user may suffer from diarrhoea. No deaths due to the ingestion of *ayahuasca* or its active ingredients have been reported.

Several Indian tribes consume *ayahuasca* during ceremonies, and medicine men drink it to seek the cause of sickness. The Tukanoan Indians of Colombia have a *Yurupari,* or initiation ceremony to admit young males into manhood. Women are excluded from participating in or even watching the ceremony. To begin this ancestor-communication ritual, the young men drink *ayahuasca,* after which they undergo a painful and bloody whipping ordeal. This is accompanied by dancing, music, and other festivities.

The *ayahuasca* ceremony typically takes place at night in a communal house. Anthropologists categorize the cere-

***Indians in the Amazon-Orinoco area of South America produce caapi by boiling the bark of the highly tangled* B. caapi. *The drink is also called* ayahuasca, *which means "vine of the souls".***

monies according to their functions, some of which are to learn the plans of an enemy; to prepare for war or hunting; to fulfil religious needs; to detect unfaithful wives; to forecast the future; to determine the cause and cure of an illness; and to experience the pleasurable effects of the hallucinogenic drink. One must realize, however, that the Indians themselves would not categorize their actions in this way. To them, the ceremonies are integrated into all aspects of the individual's and the tribe's existence, which is seen as an indivisible whole. For example, they would not separate religion from healing.

In general, the ceremony commences with the telling of legends about the origins of the particular tribe, followed by singing and dancing and the distribution of the drink at regular intervals. The hallucinogenic effects usually begin after the third or fourth cupful. The Indians believe that the brightest visions occur if during the days prior to ingesting the drug one has eaten lightly and abstained from sexual intercourse. The vividness of the hallucinations also depends upon which plants have been used in the preparation of the *ayahuasca.*

In contrast to the Indians in the forest, the urban

DR RICHARD E SCHULTES

*Leaves of* **B. inebrians,** ***another plant whose bark can produce*** **caapi,** ***which is drunk by the Indians during the*** **ayahuasca** ***ceremonies. These meetings serve many functions, such as preparation for war or hunting and forecasting the future.***

mestizos (people of mixed European and Indian ancestry) in Amazonian Peru have several night healing sessions each week. Their ceremony is much less elaborate, and the consumption of the hallucinogenic beverage is only accompanied by incantational whistling. Each participant is ritually exorcised (ridded of evil spirits) by the *ayahuasquero,* or shaman, during the ceremony, which may last up to five hours. The *ayahuasquero* is consulted by all levels of society, from merchants and professional people down to recent immigrants from the forest.

Although the *ayahuasca* prepared solely from *B. caapi* or *B. inebrians* is hallucinogenic, the addition of other hallucinogenic plants can increase or alter the effects of the drink. Two plants that are known to be added to the drink are *Diplopterys cabrerana* (formerly called *Banisteriopsis rusbyana*) and *Psychotria viridis.* Both of these plants contain N,N-dimethyltryptamine, commonly known as DMT, a highly potent hallucinogen.

The colourful hallucinations induced by *ayahuasca*

***South American natives have special names for the "diverse" species of* B. caapi, *each of which, they claim, produces different effects.***

influence much of the artwork of the Tukanoan Indians, who paint rattles, loincloths and the fronts of their houses with designs they have observed after drinking it. The colours red, yellow, and blue are most often used to adorn drawings, which often include images of snakes, panthers, jaguars, rainbows, the sun, and the male and female sexual organs. Urbanized Indians, especially mestizo medicine men in the Iquitos region of Peru, also consume the drink and incorporate their experiences in artwork.

## *Tetrapteris*

*Tetrapteris methystica* is a vine with black bark that grows in the humid tropics of the Americas. The Maku Indians of the Brazilian Amazon area use the bark to produce a yellowish drink that has hallucinogenic properties identical to those of the *ayahuasca* made from *B. caapi.* However, the hallucinogenic compounds present in *T. methystica* have not yet been identified.

**Tetrapteris methystica, *a vine with black bark that grows in the humid tropics, is used by the Maku Indians of Brazil to produce a hallucinogenic drink similar to* ayahuasca.**

DR. RICHARD E. SCHULTES

***A stone carving of Xochipili, god of inebriating flowers, discovered near the Mexican volcano Popocatepetl. The idol is adorned with stylized carvings of various flowers and vines.***

CHAPTER 6

# THE MORNING GLORY FAMILY

The morning glory family, or Convolvulaceae, is a group of climbing plants that occurs throughout the world, especially in the tropics. Within this family there are two closely related genera that contain hallucinogenic species. *Turbina,* a small genus with only 10 species, and *Ipomoea,* which has 400 species.

*T. corymbosa* (often called *Rivea corymbosa*) is a long-lived woody vine that grows from Florida down through the West Indies and Central America to the northern half of South America. It is known to be used as a hallucinogen only in Mexico. *I. violacea* is the common morning glory, an annual vine that occurs in the West Indies and from Mexico to tropical South America. It is a popular ornamental plant whose flowers come in a variety of different colours. This plant is used as a hallucinogen in Mexico and possibly also in Ecuador.

The Spanish who invaded Mexico described the Indians' use of small hallucinogenic seeds for religious and medical purposes. Though the Spaniards attempted to stop this practice, the native people continued to consume the seeds during secret ceremonies. The names of the plants from which the seeds came were not given out, but early Spanish drawings clearly show the plants to be members of Convolvulaceae, a family that modern scientists initially thought did not contain hallucinogenic species. In their records the

Spanish mentioned brown seeds called *ololiuqui* and black seeds called *tlitliltzin. Ololiuqui* are from *T. corymbosa* and *tlitliltzin* are from *I. violacea.*

*Ololiuqui* and *tlitliltzin* are still consumed by the Chinantec, Mazatec, Mixtec, and Zapotec Indians of Oaxaca, Mexico. The hallucinations produced by the seeds are used to foretell future events and to diagnose and treat various illnesses. The seeds are prepared today in the same manner described by the Spanish over 400 years ago. Despite the Indians' struggle to continue their drug rituals without being influenced by the Spanish missionaries, Christian prayers accompany the preparation of the seeds. At the beginning of a typical ceremony, the person who wishes to use the seeds' power collects them. A prayer is then offered to the Virgin Mary. On Friday night about nine o'clock, a freshly bathed virgin dressed in clean clothes grinds about a dozen seeds.

DR. RICHARD E. SCHULTES

***A flower of the Hawaiian baby woodrose,* Argyrea nervosa, *a woody, climbing vine whose seeds contain concentration of LSD-like compounds up to 15 times greater than those in morning glory seeds.***

Water is added and the mixture is then strained. The participant drinks the liquid and then lies down to allow the hallucinogens to work. By the next day future events or the source of the illness should have been revealed.

The major hallucinogenic compound in the family Convolvulaceae is *ergine,* also known as d-lysergic acid amide. Other hallucinogens, such as lysergol and ergonovine, occur in smaller amounts. The hallucinogens that occur naturally in *Turbina* and *Ipomoea* are very similar in structure to a synthetic hallucinogen called LSD, the most powerful hallucinogen known. LSD, or lysergic acid diethylamide, differs from ergine only by the presence of two ethyl ($C_2H_5$) groups. The presence of these ethyl groups makes LSD about 100 times more potent than ergine. Thus, the use of 0.05 mg (milligrams), or about two-millionths of an ounce, of LSD produces the same effect as 5 mg of ergine.

DR. RICHARD E. SCHULTES

***The* Turbina corymbosa *vine, pictured here as it grows in Colombia. The vine was a sacred plant to the Aztecs of pre-Columbian Mexico, who called it* ololiuqui *and used it for various religious and medical purposes.***

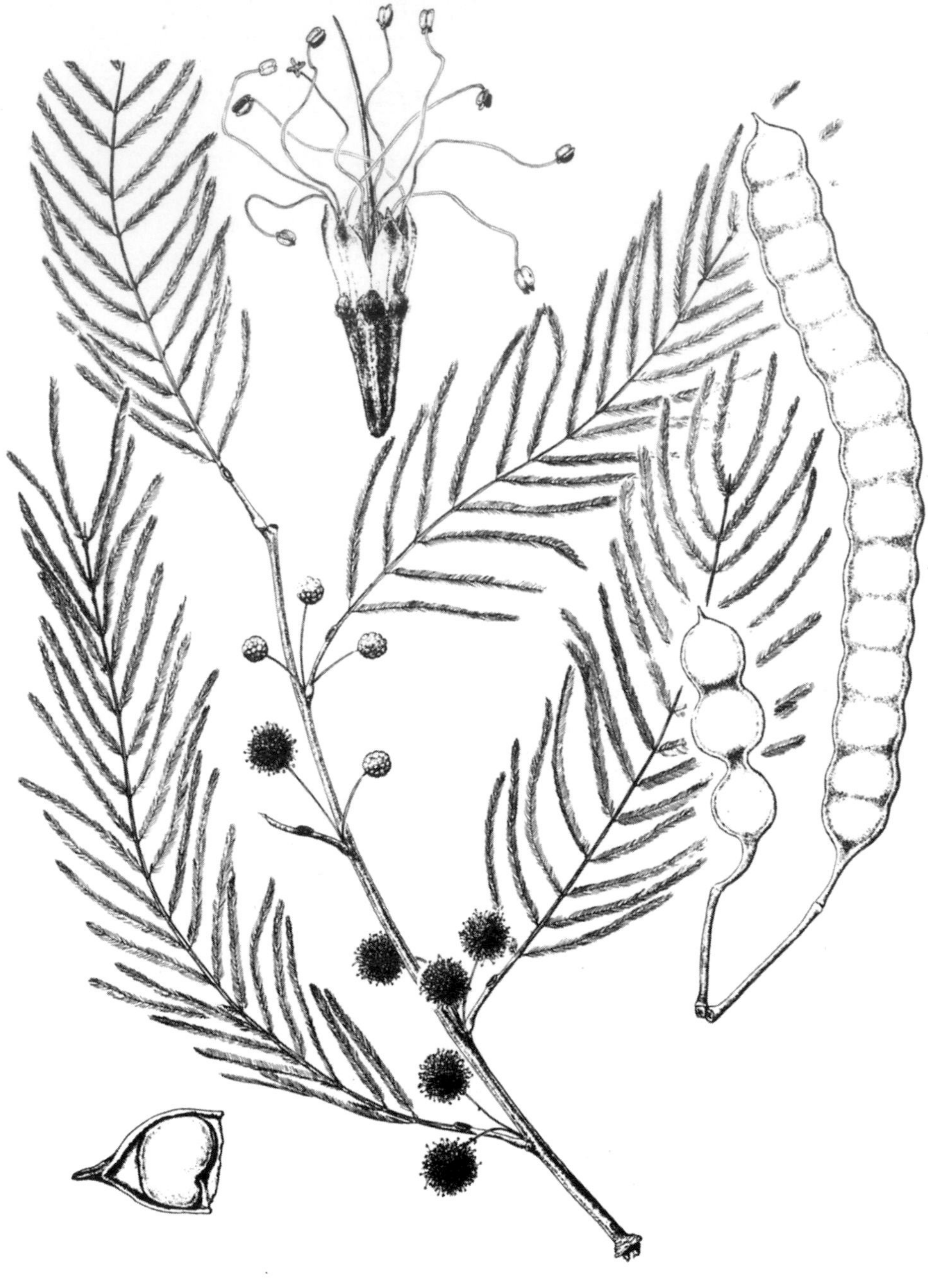

DR. RICHARD E. SCHULTES

*A drawing of the leaves, flowers, pods, and seeds of* Anadenanthera peregrina, *a tree that naturally occurs in South America and has been cultivated in the West Indies. The seeds are used by South American Indians to make a hallucinogenic snuff called yopo.*

CHAPTER 7

# THE BEAN FAMILY

The bean, or legume, family has about 14,000 species and is represented all over the world. There are two ways to classify legumes. One classification system includes them in the family Leguminosae and further divides them into three subfamilies: Mimosoideae, Caesalpinioideae and Faboideae. The second classification system includes the legumes in the order named Fabales and divides the order into three families: Mimosaceae, Caesalpiniaceae and Fabaceae. Some species of legumes contain hallucinogenic compounds.

## *Anadenanthera*

The *Anadenanthera* genus, a member of the Mimosaceae family, includes one hallucinogenic species, *A. peregrina* (formerly known as *Piptadenia peregrina*). It occurs in the West Indies and in South America and produces seeds from which a powerful hallucinogenic snuff can be prepared. In 1496 the Spanish first described how the Taino Indians of Hispaniola (an island now divided between Haiti and the Dominican Republic) inhaled the snuff to contact the spirits, and in the 19th century European explorers described its use by various Indian tribes in the Orinoco region of South America. The custom of using this hallucinogenic snuff probably originated in South America and spread to the West Indies, where the Indians planted the seeds to ensure a constant supply. Though the few surviving Indians of the West Indies no longer consume the seeds for their hallucinogenic properties, the snuff is still used by South American Indians, who call it *yopo.*

Each Indian tribe in the Orinoco region of Colombia and Venezuela prepares the snuff in a slightly different manner. The seeds are usually toasted and ground into a powder, sometimes after they are allowed to ferment. The powder is always mixed with ashes or lime derived from crushed snail shells or the bark of certain trees. *Yopo* is inhaled through tubes made from wood or from the leg bones of birds. When *yopo* is sniffed, the active chemicals are quickly absorbed by the body and produce twitching and convulsions. These symptoms are followed by nausea, hallucinations, and a deep sleep.

Another species of *Anadenanthera, A. colubrina,* grows in Peru, Bolivia, and Argentina. It has been used as a snuff known as *vilca.* An infusion can also be made from the seeds and used as an enema, a method of administration that produces hallucinations without the accompanying nausea.

Both species of *Anadenanthera* contains the same hallucinogenic compounds—tryptamines and beta-carbo-

DR. RICHARD E. SCHULTES

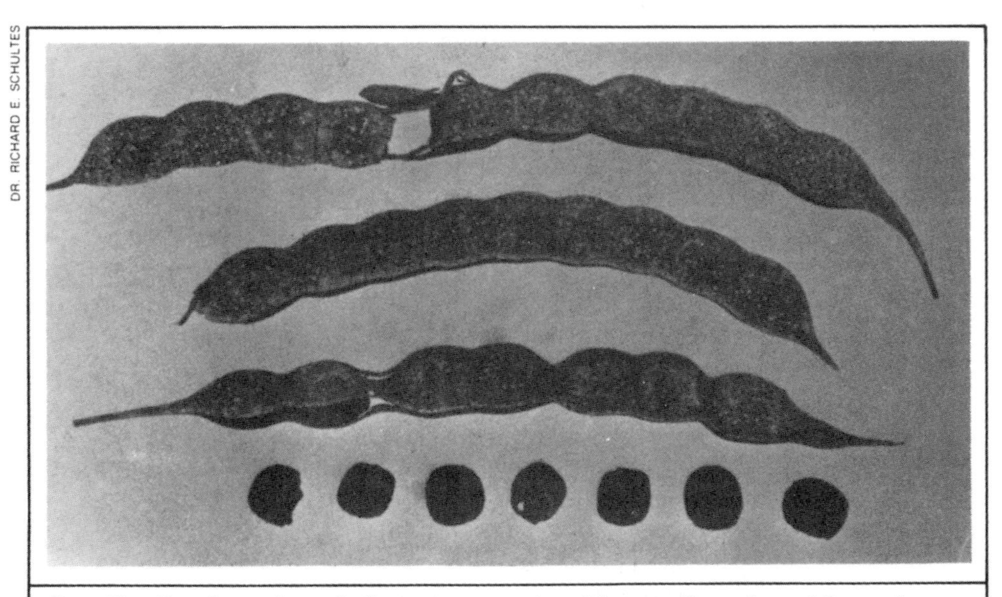

***Details of pods and seeds from* A. peregrina. *The snuff produced from the seeds, which contain a psychoactive compound related to DMT, was used by the Taino Indians to contact the spirits.***

lines—as those found in *Virola* (see Chapter 2). An interesting hallucinogenic compound called *bufotenin* has also been isolated from *Anadenanthera.* Bufotenin is also found in the skin secretions of toads and in some mushrooms.

## *Mimosa*

The genus *Mimosa,* another member within the Mimosaccae family, contains about 500 species. The roots of *M. hostilis* was used in eastern Brazil as a source of the hallucinogens included in a drink called *juremo.* The tribes that prepared and drank *juremo* before going into battle are now extinct, so the exact methods of preparing the liquid will never be known. The roots are known to contain the powerful hallucinogen DMT.

DR. RICHARD E. SCHULTES

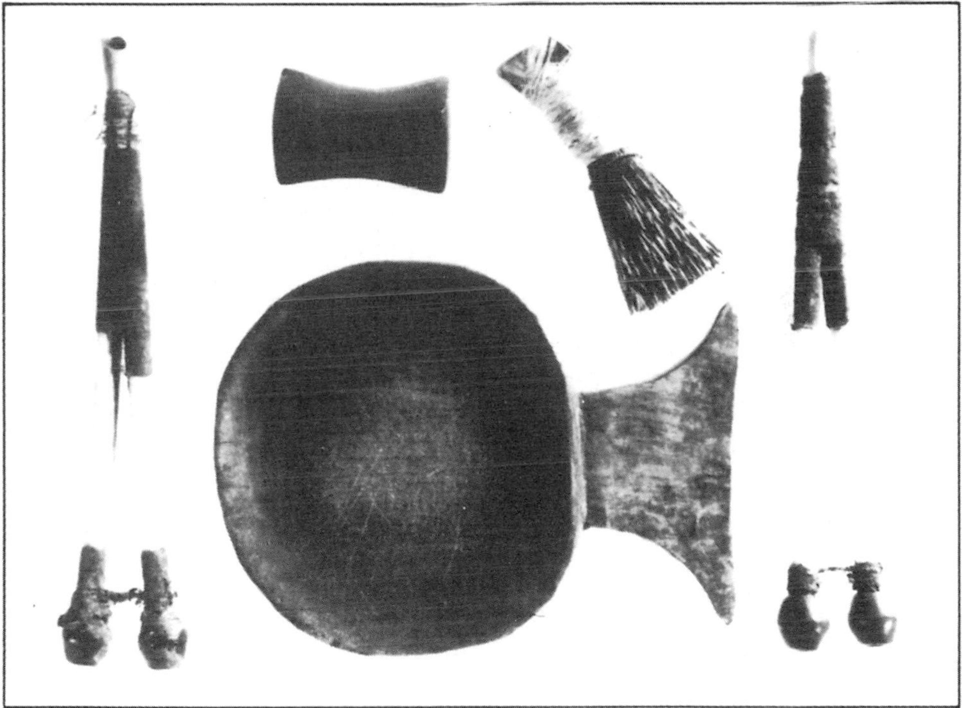

***Snuff tubes and a mortar and pestle for the use and preparation of yopo, which, when sniffed, can produce convulsions.***

### *Cytisus and Sophora*

The Fabaceae family has two hallucinogenic legumes that contain compounds called *quinolizidine alkaloids.* These plants are *Cytisus canariensis* and *Sophora secundiflora.*

*C. canariensis* comes from the Canary Islands, which are on the northwestern coast of Africa, though it has also been introduced into Mexico. This plant is the "genista" found in florists' shops. It is employed by the medicine men of the Yaqui tribe of northern Mexico, though it is rare for an imported plant to be used by medicine men in the New World. The genus *Cytisus* usually contains *cytisine,* a quinolizidine alkaloid. In addition to inducing hallucinations, cytisine can be toxic and cause nausea, convulsions, and death due to respiratory failure.

The genus *Sophora* occurs in warm regions of the Old and New Worlds. *S. secundiflora* is the source of hallucinogenic mescal beans. The plant is a small tree that grows in northern Mexico and in neighbouring Texas and New Mexico. It has blue flowers that turn into silvery pods containing up to eight or more bright red seeds. The seeds contain the quinolizidine alkaloids *cytisine* and *sparteine,* both of which are thought to produce hallucinations. These substances can also cause nausea, convulsions, and death from asphyxiation (a lack of oxygen) due to their depressive effect on the diaphragm.

The Indians who lived in the area of the mescal-bean tree used the beans as a medicine and as a hallucinogen. Large quantities of beans have been found in excavations dated 7500 BC. Since the beans cannot be used as food, it may be assumed that they were used for ceremonial purposes. The Red Bean Dance, which utilized the beans as an oracular, divinatory, and hallucinatory medium, was performed by a number of Indian tribes in the United States such as the Apache, Comanche, Delaware, Iowa, Kansa, Omaha, Oto, Osage, Pawnee, Ponca, Tonkawa, and Wichita tribes.

At the end of the 19th century, the Red Bean Dance disappeared as the mescal bean was replaced by peyote in Indian ceremonies (see Chapter 4). Peyote produces more colourful visions than those caused by the mescal bean. In addition, peyote is much safer. One must be extremely cautious when consuming the mescal bean, since even a small overdose can cause nausea and death by asphyxiation. The mescal bean must have been very important to the Indians if they used it for thousands of years knowing that it can so easily lead to death.

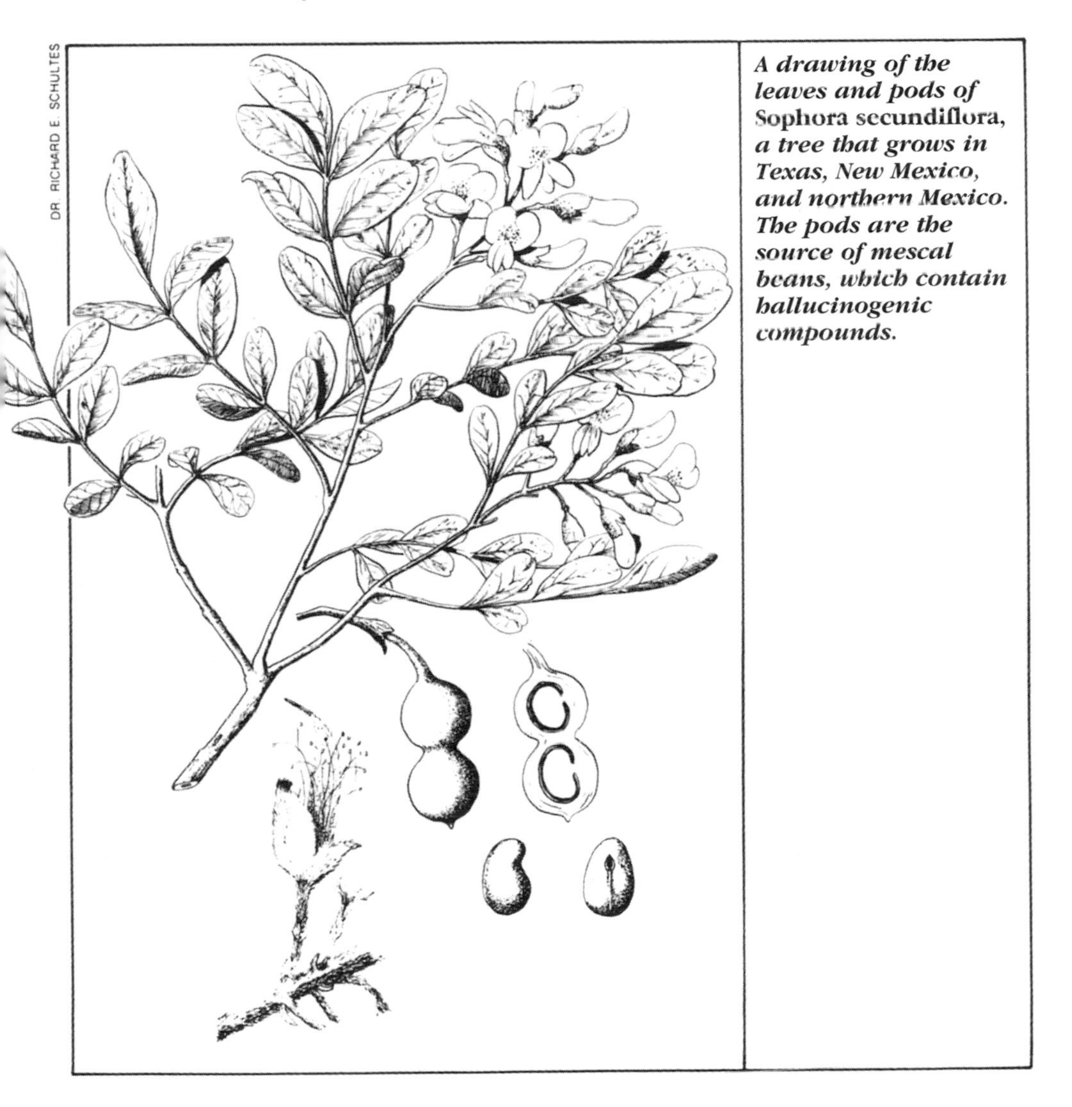


***A drawing of the leaves and pods of* Sophora secundiflora, *a tree that grows in Texas, New Mexico, and northern Mexico. The pods are the source of mescal beans, which contain hallucinogenic compounds.***

DR. RICHARD E. SCHULTES

**Atropa belladonna,** *or the deadly nightshade, contains the psychoactive drugs* **hyoscyamine** *and* **scopolamine,** *which produce hallucinations and can cause headaches, increased heart rate and even death.*

CHAPTER 8

# THE POTATO FAMILY

The potato family, or Solanaceae, is a large family that occurs throughout most of the world and includes such well-known plants as the potato (*Solanum tuberosum*), the tomato (*Lycopersicon esculentum*), and the tobacco plant (*Nicotiana tabacum*). The plants in this family commonly produce poisonous alkaloids. Potatoes exposed to light prior to harvesting turn green, a sign of the presence of a toxic alkaloid called *solanine.* If ingested, solanine can cause symptoms such as headaches, vomiting, diarrhoea, fever, apathy, restlessness, confusion and hallucinations. Other plants in the potato family produce compounds called *tropanes,* which are responsible for the hallucinogenic properties of solanaceous plants from the Old and New Worlds. Plants that contain tropanes include species of *Atropa* (belladonna, or the deadly nightshade), *Hyoscyamus* (henbane), *Mandragora* (mandrake) and *Daturu* (thorn apples). *Daturu* grows in both hemispheres, though the other plants are native to the Old World.

The hallucinogenic species of Solanaceae from the Old World are often called the hexing herbs because they were used in the green ointments, or flying potions, of European witches. A typical flying potion contained many substances, usually including *Atropa belladonna, Datura stramonium, Hyoscyamus niger, Mandragora officinarum,* and the fat of a child. The last ingredient helps to explain why witches were so disliked—they dug up dead babies and boiled them in pots to obtain the fat from their bodies. This mixture of plants and fat formed a green ointment and was used to induce "flying". The ointment was rubbed on the body, and the fat helped the tropanes to pass through the skin and enter the bloodstream. The process could be speeded up by applying the ointment to the genitals (male or female) and/or the anus, where the rich supply of blood vessels readily absorbed the hallucinogenic compounds.

Sometimes the witches used a staff made of ash wood to apply the ointment to themselves, but because witchcraft was against the law (frequently carrying the death penalty), they often disguised the staff as a broomstick. This practice led to the stories of witches flying on broomsticks. Of course, the ointments did not actually cause the witches to fly, but instead quickly produced a deep, dream-filled sleep. Many of the dreams included the sensation of flying through the air as well as visions of dancing and having sexual orgies with the

THE BETTMANN ARCHIVE

*A 15th-century print of a woodcut from* **De Lamiis et Phitonicus Mulieribus** *by Molitor, depicts a wizard and two witches, with the heads of an ass and a hawk, riding on a forked branch to the Witches' Sabbath. The branch was said to be a tool used by witches when they prepared the ointments that allowed them to fly. These potions often contained parts of the mandrake plant, which was believed to have certain magical properties.*

devil. Recent scientific experiments have confirmed that the hallucinogenic tropanes, such as those found in the potato family, do indeed produce these types of dreams as well as dreams involving *lycanthropy,* the transformation of humans into wolves or other predatory animals.

## *The Deadly Nightshade*

*Atropa belladonna* is a poisonous perennial herb that is native to Europe but is now cultivated in many countries as a source of useful medicines. It is called the deadly nightshade because ingestion of its attractive, shiny, blue-black berries is fatal. In the Middle Ages, Italian ladies used to drop *Atropa* sap into their eyes to produce a glassy stare, which was believed to enhance their beauty. This practice gave rise to the other common name for this plant, *belladonna,* which in Italian means "beautiful lady".

The major hallucinogenic compound in *A. belladonna* is *hyoscyamine. Scopolamine,* another hallucinogen, is present in smaller quantities but is more potent than hyoscyamine.

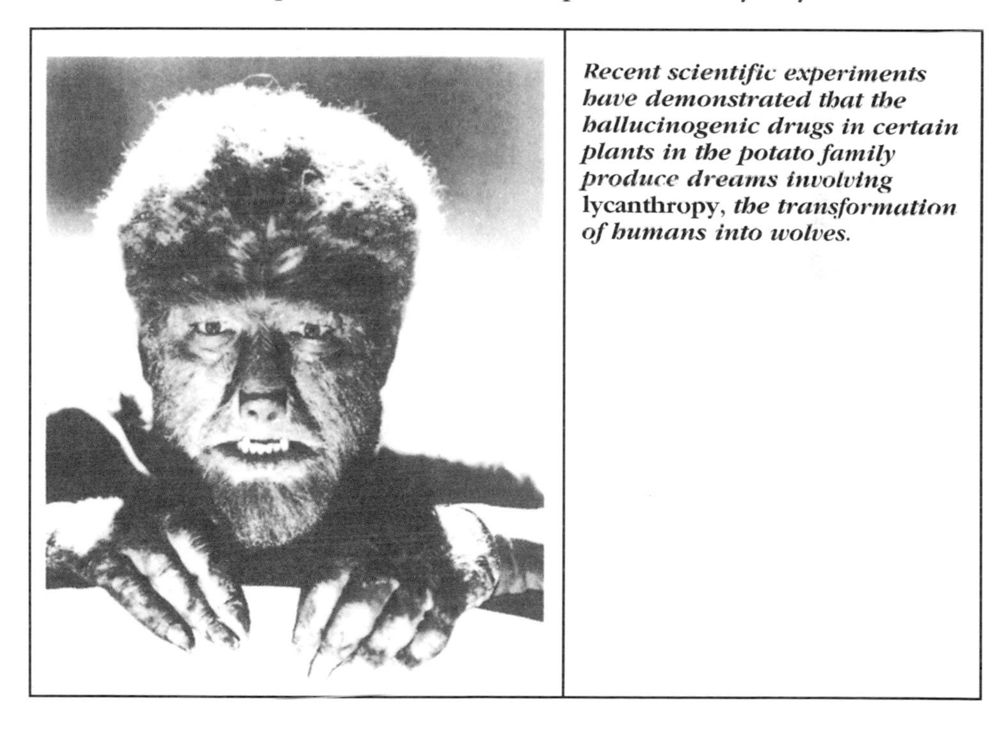

***Recent scientific experiments have demonstrated that the hallucinogenic drugs in certain plants in the potato family produce dreams involving* lycanthropy, *the transformation of humans into wolves.***

These compounds are present in all parts of the plant, with the greatest concentrations occurring in the leaves, roots, and seeds. The hallucinations produced by both substances are followed by a deep sleep.

### *Henbane*

*Hyoscyamus niger* is an annual or biennial plant that was native to Europe but is now found across Asia and North America. Commonly known as henbane ("hen's death" in Middle English), it contains scopolamine and hyoscyamine, and thus is extremely poisonous. Small doses will produce sleep and hallucinations, but a large dose can be fatal. In the Middle Ages henbane was cultivated around the Bohemian city of Pilsen, and the crushed seeds were added to beer to make it more intoxicating. Today the leaves and seeds of the henbane plant are used for the production of some sedatives.

### *Mandrake*

*Mandragora officinarum,* or the mandrake plant, is a perennial that has a long root, large leaves, and bluish flowers that give rise to yellow fruits. It is native to the area around

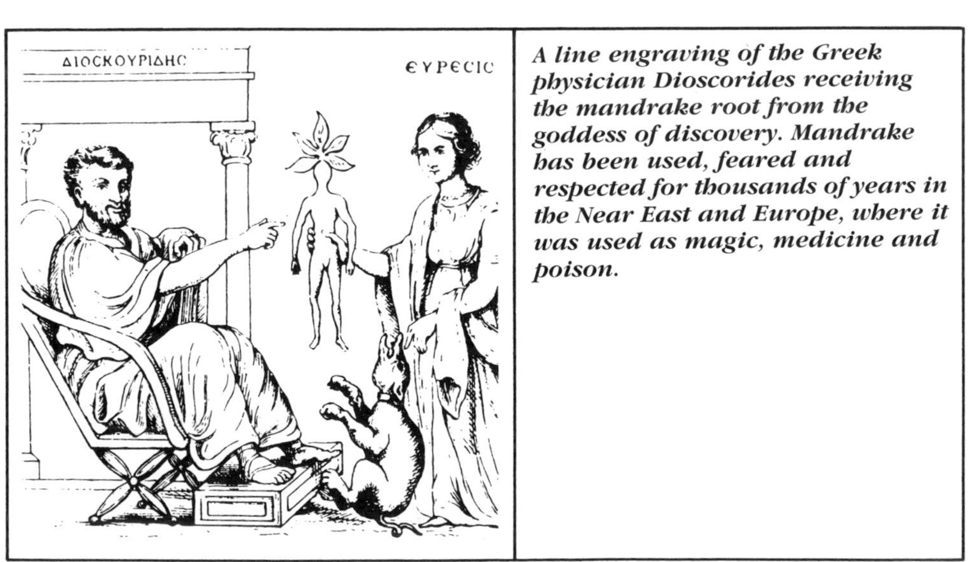


THE BETTMANN ARCHIVE

***A line engraving of the Greek physician Dioscorides receiving the mandrake root from the goddess of discovery. Mandrake has been used, feared and respected for thousands of years in the Near East and Europe, where it was used as magic, medicine and poison.***

the Mediterranean Sea but has been cultivated throughout Europe and even in the New World. Hyoscyamine and, in lesser amounts, scopolamine and atropine are the major hallucinogenic compounds present in the plant.

The mandrake plant has an unequalled place in folklore because its root often resembles a human body. People thought that such a plant must surely be endowed with all kinds of magic properties. The Greek writer Isidorus (AD 500) and the Egyptian writer Serapion (AD 350) both described the use of mandrake as a painkiller for surgical operations. Mandrake was also given to victims of crucifixion to cause unconsciousness and thus eliminate the pain and agony of their deaths. In the Middle Ages the use of mandrake was dictated by the "Doctrine of Signatures", which stated that the appropriate plant to be used for treatment was ordained by the part of the body that the plant resembled. Mandrake roots whose shape suggested the male body were prescribed for male-oriented diseases, and the more slender, female-looking roots were prescribed for female-oriented problems. Some people claimed that mandrake extracts had remarkable powers of rejuvenation, similar to those claimed for ginseng by the Chinese. The root was also included in love potions, and the fruits were said to increase fertility. One legend claims that the mandrake grew under the gallows



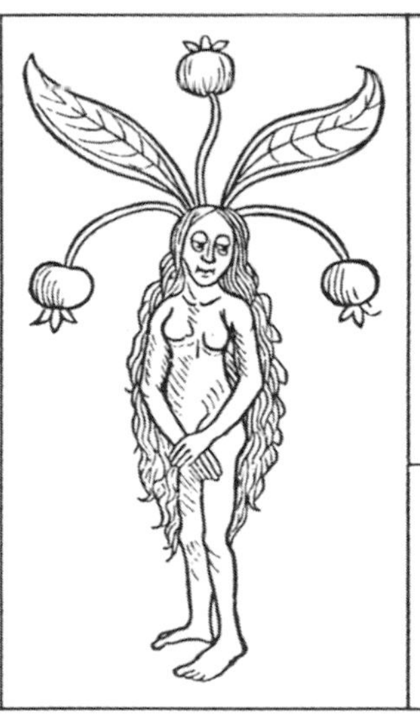

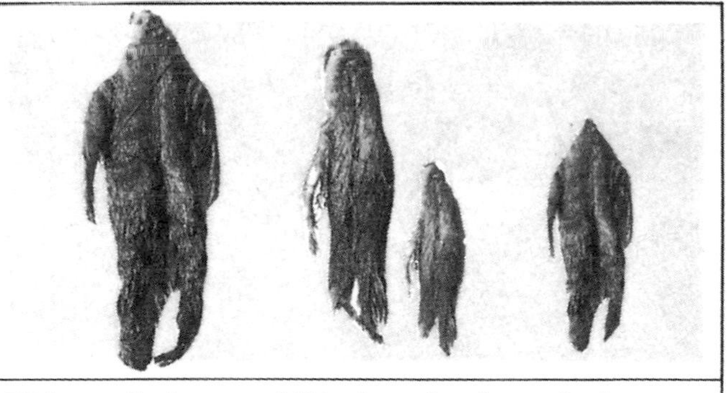

***FDA regulations prohibit the sale of mandrake (above) for human consumption. The plant contains significant amounts of potentially dangerous hallucinogens. At left is a woodcut of an anthropomorphized mandrake root, from an ancient textbook of medical science.***

used to hang criminals, whose blood, urine, or semen gave rise to the plant. These uses and legends have never been scientifically proven, but the plant was both respected and feared. In some areas of Europe, possession of the root was punishable by death.

The root of the plant was taken from the ground with the greatest of caution. As early as AD 93 the historian Flavius Josephus (c. AD 37-100) described the process, stories of which were embellished over the years. Many people believed that the mandrake shrieked when harvested and that anyone hearing the piercing cry would die. To avoid this, dogs were used to gather the root. The dog was starved for several days and then tied to the root, around which a trench had been cut. The owner stood out of earshot and threw a piece of meat, and as the dog leapt for the meat, the mandrake root was pulled from the ground. Some writers

THE BETTMANN ARCHIVE

***Lucrezia Borgia (1480-1519), in a painting by Bartholomeo da Venzia. Italian ladies sometimes dropped sap from Atropa into their eyes, producing a glassy stare that they thought enhanced their beauty. This practice gave rise to the plant's common name, belladonna, or "beautiful lady".***

actually stated that the dog immediately died. There are also references to the use of a sword to draw three circles around the plant and to the fact that the plant could be removed only after sundown.

## *Datura*

Hallucinogenic species of *Datura,* or thorn apples (a term that refers to the shape and appearance of the fruits), occur in both the Old and New Worlds, where they have a long history of use as medicines and hallucinogens. Noted specifically for their hallucinogenic qualities are the Old World species *D. fastuosa,* and *D. metel,* and the New World

Datura, ***a sacred plant in Afghanistan, is mentioned in early Arabic, Chinese and Sanskrit writings. Only alcohol has been in use for as long and over as large a part of the world as*** **Datura.**

species *D. stramonium,* or jimsonweed. The name *jimsonweed* is an abbreviation of *Jamestown weed.* In the 17th century, British soldiers in Jamestown, Virginia, picked the plant and ate its leaves for food. They became deranged for several days and were jailed to keep them from injuring themselves. After 11 days they recovered and could remember nothing of the incident. *D. metel,* which is mentioned in early Arabic, Chinese, and Sanskrit writings, is native to Asia, though it is now also grown in Africa and America. *D. ferox* is originally from Asia, where it has been consumed for its medicinal and hallucinogenic properties. All the hallucinogenic species of *Datura* contain tropanes. Hyoscyamine and, in lower concentrations, scopolamine are generally present.

Species of *Datura* have been used to treat such ailments as pneumonia, heart disease, and hysteria. The sleep caused by *Datura* may last up to 20 days. In some parts of Asia, the plant has been added to food or tobacco and used by thieves to put their victims into a deep sleep. Similarly, it has been

DR. RICHARD E. SCHULTES

***An Afghan man with* Datura metel. *Some species of* Datura *induce a sleep that can last for as long as 20 days.***

used to lure young girls into prostitution. In India *Datura* is sometimes mixed with cannabis and smoked. High doses of the plant can cause death.

During the female puberty rites of the Tsonga tribe of Mozambique and South Africa, *D. fastuosa* is mixed with fat and rubbed on the bodies of the young women. Similar initiation rites exist in the New World. The Algonquin Indians of eastern North America used a medicine called *wysoccan,* whose main hallucinogenic ingredient was *D. stramonium.* The young men being initiated were kept in seclusion for many days, after which they were given the *wysoccan.* They were kept in a state of violent intoxication for almost three weeks, during which time they were expected to forget the memories of their childhood. If childhood memories were found found to be retained, the unfortunate young men were given a second and larger dose of the medicine, which ocassionally resulted in death.

The dried roots of *D. inoxia* (formerly known as *D.*

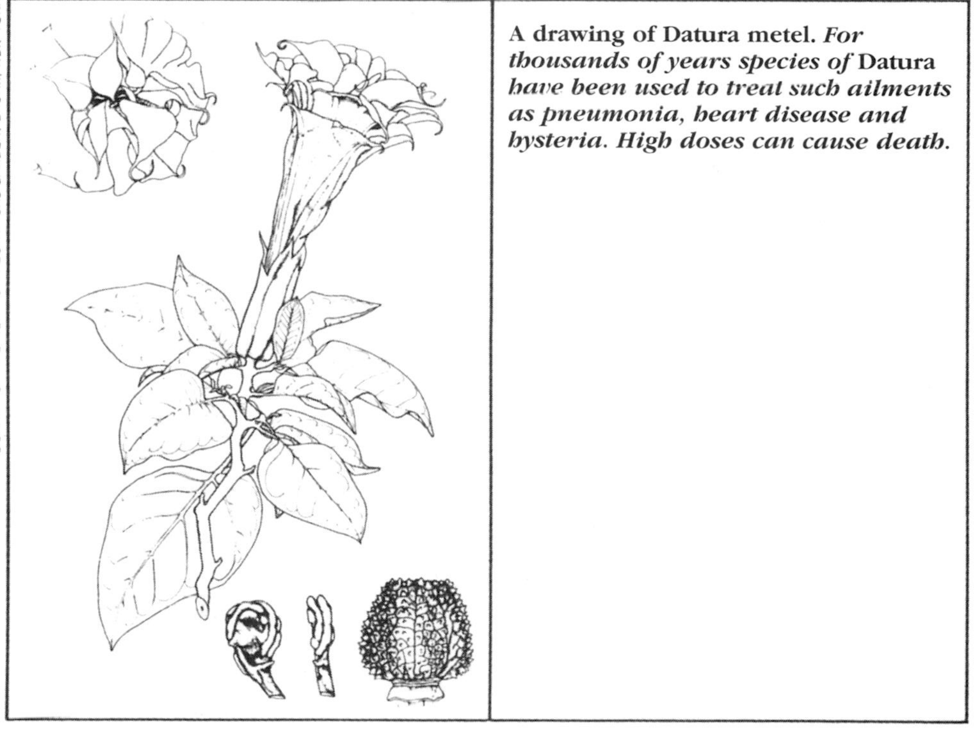

LIBRARY OF THE NEW YORK BOTANICAL GARDEN, BRONX, NEW YORK

**A drawing of Datura metel.** ***For thousands of years species of*** **Datura** ***have been used to treat such ailments as pneumonia, heart disease and hysteria. High doses can cause death.***

*meteloides*) were an integral part of initiation rites in Arizona, California, and New Mexico. These rites were usually limited to young males, but in some tribes both males and females were initiated. After a period of instruction and fasting, *toloache,* a hot water extract of the root, was given to each initiate. This was immediately followed by ceremonial processions and dances until the participants fell into an unconscious trance, which sometimes lasted up to three nights. During this time the initiates had many visions, often of animals. However, some young people never awoke from the trance and died from an overdose of the tropanes. Recently several people have died in California after trying to imitate these ancient ceremonies in order to experience the *toloache* hallucinations.

The rain priests of the Zuni Indians of New Mexico used to chew the roots of *D. inoxia* to communicate with the spirits of the dead so that they would petition the gods for rain. The Zuni also used the plant as an anaesthetic and as a treatment for wounds. The Yuman Indians consumed *D. inoxia* to gain the ability to foretell the future.

The greatest use of *Datura* occurs in Mexico. The Spanish described the medicinal application of this plant in Mexico several hundred years ago and noted that madness resulted from consuming excessive amounts. The Tarahumara Indians still add *D. inoxia* to maize beer to make it stronger and hold ceremonies in which hallucinogenic visions are brought on by consumption of the plant. However, many Mexican Indians prefer the use of the safer peyote cactus over *Datura.* In fact, in Mexico *D. ceratocaula* is known as *torna-loca*—the plant that makes you crazy.

### *Other Hallucinogenic Genera*

There is an interesting group of plants that was formerly considered to be within the genus *Datura,* but which is now placed in a separate genus: *Brugmansia.* This genus contains five or six species of small trees, all of which are native to South America. Often called tree-daturas, *Brugmansia* plants do not occur in the wild but are cultivated by various Indian tribes, who grow the plants from pieces of stem. The plants are used for their medicinal and hallucinatory properties just as *Datura* is. To use the plants as a hallucinogen, either a tea is made from the leaves or the seeds are added to alcoholic drinks or to the beverage prepared from *Banisteriopsis* (see Chapter 5). *Brugmansia* plants contain the same tropanes that are found in *Datura.*

Several other hallucinogenic genera in the Solanaceae family occur in South America. These include *Iochroma; Brunfelsia,* which the Indians call *borracher,* or "intoxicator"; *Latua,* referred to as *arbol de los brujos,* or "the sorcerers' tree" and *Methysticondendron,* or *mitskway borrachero* ("snake intoxicant"). The plants within this genera were used or are still used by various Indian tribes as sources of hallucinogens.

LIBRARY OF THE NEW YORK BOTANICAL GARDEN, BRONX, NEW YORK

***In 1934 the Burgess Seed & Plant Company used this advertisement to sell* Brugmansia, *"a magnificant indoor and garden plant". Purchasers were undoubtedly unaware of the hallucinogenic properties of the seeds they received in the mail.***

DR. RICHARD E. SCHULTES

*A drawing of* Justicia pectoralis, *a hallucinogenic herb that grows in the West Indies and in Central and South America.*

# CHAPTER 9

# OTHER HALLUCINOGENIC PLANTS

In many plant families there are only one or two hallucinogenic species. The hallucinogenic compounds in some of these plants have not yet been identified or confirmed by scientific testing, but new discoveries are made each year.

### *Sinicuichi*

The loosestrife family, or Lythraceae, consists mainly of tropical herbs. Within this family there is a shrub called *Heimia salicifolia,* or sinicuichi, which grows in the West Indies and from Texas to northern South America. Six hallucinogenic quinolizidine alkaloids, of which *cryogenine* is the most active, are present in this plant. To prepare sinicuichi for consumption, the leaves are picked, allowed to wilt, and then crushed in water, after which the mixture is put in the sun to ferment. Drinking the resulting liquid causes giddiness, drowsiness or euphoria, and auditory hallucinations or deafness. Some Mexican Indians believe that the consumption of this plant helps them remember past events, including those that occurred before they were born. Excessive consumption is said to be harmful.

## *Desfontainia*

The genus *Desfontainia* is placed either in its own family, Desfontainiaceae, or in a larger family called Loganiaceae. *D. spinosa* is a shrub that grows in the highlands from Costa Rica to Chile. Some Indian tribes of Colombia and the Mapuche Indians of Chile make a hallucinogenic tea from the leaves. The medicine men drink the tea to give them the ability to foretell the future and diagnose disease. The hallucinogenic compounds in this plant have not yet been identified, though it is known that other members of Loganiaceae produce alkaloids containing an indole ring.

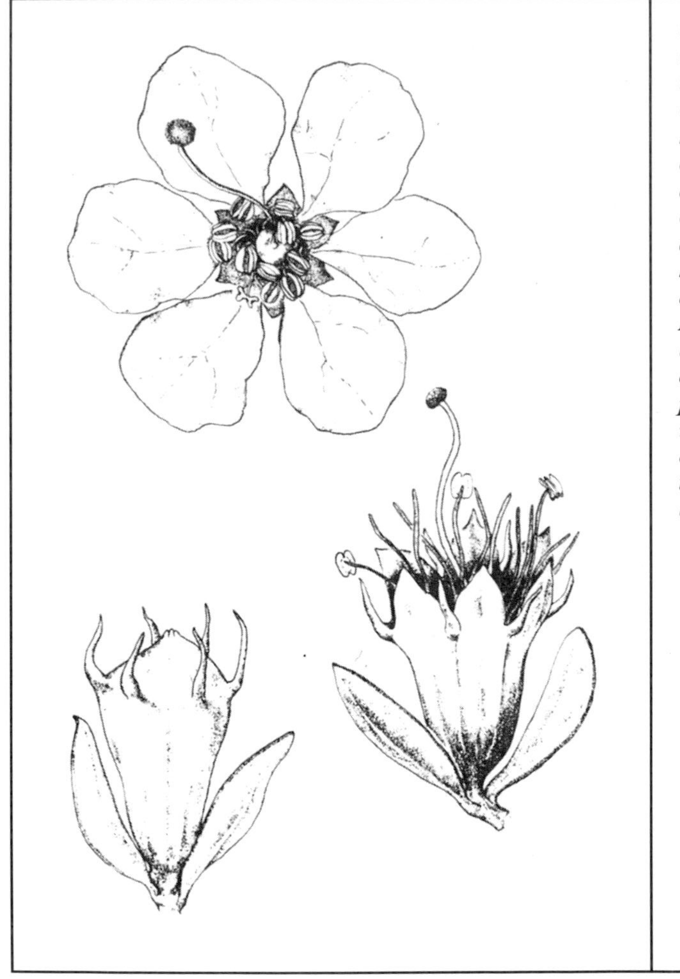

**Sinicuichi,** ***or*** **Heimia salicifolia,** ***is a shrub that contains*** **cryogenine,** ***a hallucinogen that causes giddiness, drowsiness or euphoria, and auditory hallucinations or deafness. Some Mexican Indians believe that consumption of this plant helps them remember past events, even those that occurred before birth.***

### *The Dogbane Family*

The dogbane family, or Apocynaceae, occurs widely in the tropics. (In Middle English, dogbane means "dog killer".) The family is well known for its alkaloids, some of which are now used to cure certain types of Leukaemia. At least two species of the dogbane family contain hallucinogenic alkaloids with an indole ring. One of them, *Apocynum quebrancho-blanco,* is a New World plant containing the compound *yohimbine.* When brewed as a tea and drunk, this substance can cause mild hallucinations and a heightening of emotional and sexual feelings as well as an increase in blood pressure.

DR RICHARD E SCHULTES

***A member of the dogbune family,*** **Tabernanthe iboga** ***contains the psychoactive drug*** **ibogaine,** ***a hallucinogen, stimulent and aphrodisiac. In Gabon, the roots are used in the initiation rites of secret societies such as the Bwiti cult.***

Anyone with diabetes, kidney, and/or heart disease should not experiment with this drug. In addition, in combination with certain foods, such as chocolate, cheese, bananas, and pineapples, yohimbine can cause dangerously high blood pressure accompanied by shortness of breath. Adverse reactions must be treated immediately by a medical professional.

The second hallucinogenic species of the dogbane family is *Tabernanthe iboga.* This shrub is native to Gabon and the Congo in West Africa, though it is also cultivated in Gabon, which provides a constant supply of the yellow root.

***A Waiká Indian of the Brazilian-Venezuelan frontier holding a basket of* Justicia *leaves, which are prepared for consumption by being dried and then ground into a fine powder. Sometimes the powder is added to the snuff made from* Virola *resin.***

The active hallucinogenic ingredient is *ibogaine*. The use of the plant was first reported by French and Belgian explorers in the 19th century. *T. iboga* forms the basis of the Bwiti cult, which has successfully resisted Christian and Islamic missionaries. The cult continues to grow, using the drug during night time ceremonies, which include dancing and drumming. Small doses act as a stimulant and an aphrodisiac, but larger amounts can cause death.

### *The Mint Family*

The mint family, or Labiatae, occurs throughout the world. Most of the species in this family are herbs, though there are a few climbing plants and small trees. The family is rich in volatile or essential oils (plant oils that contain the plant's characteristic perfumes and medicinal compounds) and many species are used extensively in herbal medicine. *Salvia* is the largest genus and contains 500 species. *S. divinorum,* the only species known to have hallucinogenic properties, exists as a cultivated plant in Oaxaca, Mexico. The Mazatec Indians of this area consume this plant to enable them to foresee the future, which explains the plant's scientific name—"the salvia of the diviners". The Mazatec refer to *S. divinorum* as *hojas de la Pastora* or *shka-Pastora,* both of which mean "leaves of the Shepherdess". There are two ways to consume this hallucinogenic plant: the leaves can be picked and chewed, or they can be ground on a *metate* (grinding stone), mixed with water, and then strained and drunk. Though the hallucinogenic compounds in *S. divinorum* have not yet been isolated and identified, it is known that consumption of the plant produces short-term visual hallucinations, which often include three-dimensional designs in kaleidoscopic colours.

### *The Acanthus Family*

The acanthus family, or Acanthaceae, consists mainly of tropical herbs and shrubs. *Justicia,* the largest genus, has approximately 300 species, only one of which is known to have hallucinogenic properties. *J. pectoralis* grows in the West Indies and in Central and South America and is used as a hallucinogen in Colombia and Brazil. The Waiká Indians of the Brazilian-Venezuelan frontier call the plant *masha-hari* and grow it around their houses. The leaves are prepared for consumption by drying them and then grinding them into a powder. The powder may be sniffed or added to the hallucinogenic snuff prepared from *Virola* resin (see Chapter 2). There is some evidence that the leaves contain the powerful visual hallucinogen DMT.

## The Madder Family

The madder family, or Rubiaceae, consists mainly of tropical trees, shrubs, and woody vines of the Old and New Worlds. Coffee is a member of this family. At least two genera contain hallucinogenic species. One is *Pausinstalia. P. yohimba,* native to Cameroon and Congo in Africa, contains the indole alkaloid yohimbine (the same compound found in *Apocynum quebrancho-blanco*). Yohimbine causes visual hallucinations, but it is not known if *P. yohimba* is used as a source of hallucinogens.

*Psychotria* is the largest genus in the madder family. In South America the leaves of several *Psychotria* species are used as additives to the hallucinogenic beverages prepared from *Banisteriopsis* (see Chapter 5). It is interesting to note that several tribes of Indians from widely separated areas in Brazil, Colombia, Ecuador, and Peru use *Psychotria* leaves in this way. *P. viridis* is usually the plant chosen for this purpose, but at least four other species are known to be used as well. These plants contain DMT, a hallucinogen found in many other plants.

## The Aster Family

The aster family, or Compositae, is a very large family of flowering plants. Many different types of chemical compounds are found in this family, and thus it is not surprising that some of these are hallucinogenic. Indian tribes in Mexico use two New World species of Compositae to produce hallucinogenic visions: *Calea zacatechichi* and *Tagetes lucida.* The hallucinogenic compounds in these species have not yet been identified.

The Chontal Indians of Oaxaca dry the *Calea zacatechichi* leaves and then smoke them in cigarettes and/or make a hallucinatory tea, which is slowly sipped. This is said to clear the senses and to facilitate communication with the spirit world. Used in folk medicine to treat fevers, nausea, and other complaints, the leaves produce drowsiness followed by hallucinations. There is apparently no cult or religion associated with the use of this plant.

The leaves of *Tagetes lucida* are also smoked, either alone or mixed with tobacco. The Huichol Indians claim that smoking the plant during ceremonies produces visions and hallucinations.

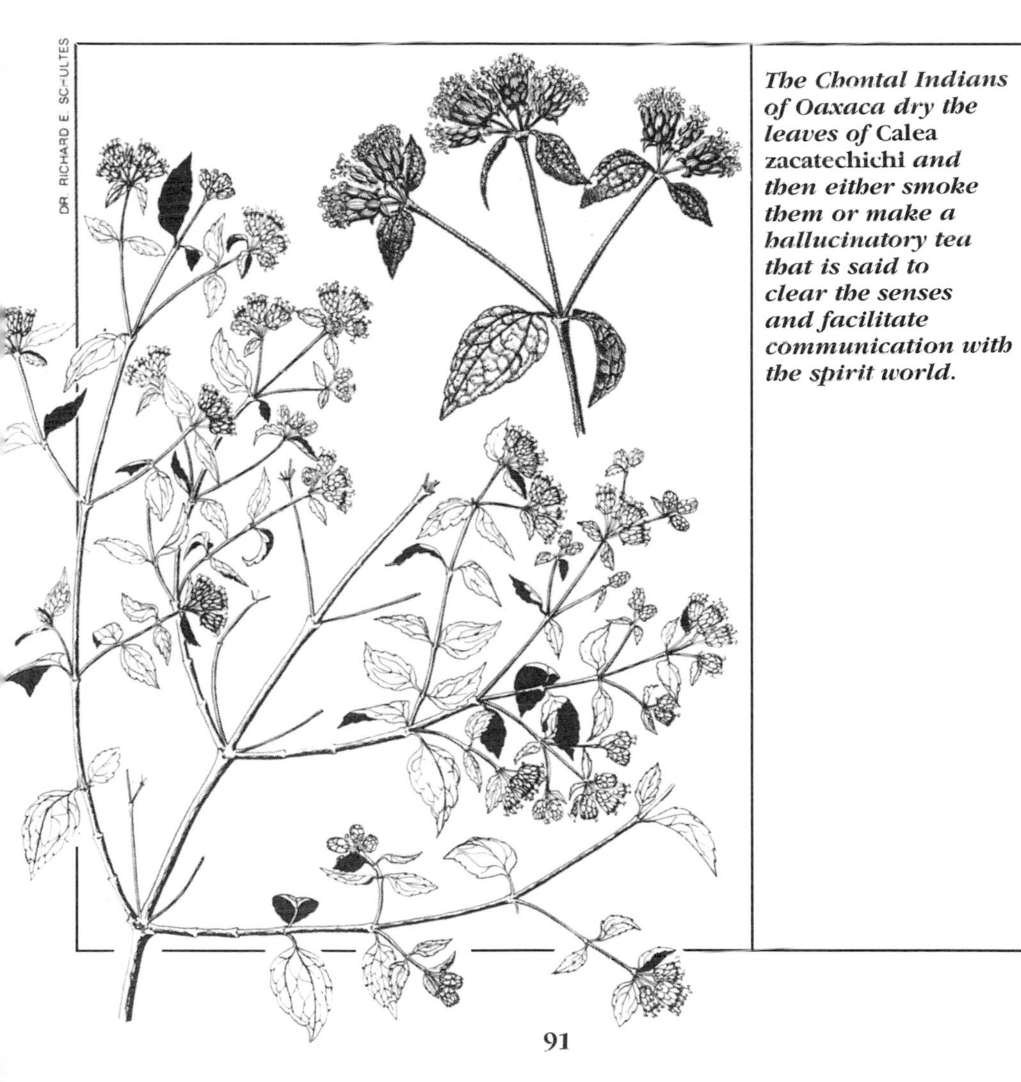

DR. RICHARD E. SCHULTES

***The Chontal Indians of Oaxaca dry the leaves of* Calea zacatechichi *and then either smoke them or make a hallucinatory tea that is said to clear the senses and facilitate communication with the spirit world.***

*The opium poppy,* Papaver somniferum, *was mentioned over 4,000 years ago by the Sumerians, who recognized the plant's ability to induce sleep and relieve pain. It also appeared in ancient Greek and Roman literature.*

## CHAPTER 10

# OPIUM AND COCAINE

There are two well-known psychoactive plants that are not hallucinogenic: the opium poppy and the coca plant. They fall into a different group of plants that contain chemical compounds that are classified as analgesics and euphorics. An analgesic, such as that in the opium poppy, causes an insensitivity to pain without loss of consciousness, and a euphoric, such as that in the coca plant, causes a feeling of general well-being.

### *The Opium Poppy*

The opium poppy, native to Europe and Asia, is in the Papaveraceae family and has the scientific name *Papaver somniferum* ("the poppy that brings sleep"). Almost 4,000 years ago the Sumerians who lived in ancient Babylonia (modern Iraq), mentioned that consumption of the plant brought sleep and an end to pain. The ancient Greeks and Romans often mentioned this plant in their literature. After the decline of the Roman Empire, the use of the opium poppy spread through the Islamic Empire and reached India and Southeast Asia.

The Chinese grew *P. somniferum* as a garden plant rather than as a source of opium. In the 19th century the English authorities in India supervised the cultivation of huge amounts of the opium poppy, and they chose China as the market for the opium produced. At this time China was the sole source of tea, the demand for which was increasing in the Western world. China was essentially self-sufficient but gladly accepted silver in exchange for the tea. The English began to export opium to China and, as the number of opium users became larger and larger, the silver exchanged for opium outgrew the silver given to the Chinese for tea. This caused a serious imbalance of payments, and the ruling Manchu government passed laws that prohibited the importation of opium. These laws annoyed the English, who had begun to rely on the surplus of silver to finance trade in the Far East. The English were also anxious to open up the vast potential Chinese market for manufactured goods, especially cotton clothing. Using their vastly superior ships and cannons, the English fought and won the First Opium War (1839–1847), opening up China for trade, and the Second Opium War (1856–1860), forcing the Chinese to legalize the importation of opium.

Thus, the English turned China into a nation of opium addicts, taking valuable silver and tea in exchange for the dangerous opium. However, in 1907 Great Britain, China, and India agreed to cease the cultivation and trading of opium over a 10-year period and, as a result, the opium

***A steel engraving depicts the capture of Chin-Keang-Foo, during a battle in the First Opium War (1842). The conflict involved the Chinese and the British, who objected to China's ban on the importation of opium.***

problem in China virtually disappeared.

In the 19th century opium was legal, freely available, and widely used in Britain and the United States. It was even given to children to promote sleep. At the time, opium was a cheap substitute for alcohol, and some doctors encouraged its use for this purpose. As recently as 1940, doctors in Kentucky were converting alcoholics into morphine users. (Morphine is the principal psychoactive ingredient in opium.)

The flowers of *P. somniferum* produce a green, seed-containing capsule. If the side wall of the capsule is cut, a milky latex is released—opium. When allowed to dry, the opium becomes a brown gum that can be peeled easily from the capsule. In the past, people smoked this raw opium, though today the gum, which contains 24 alkaloids, including morphine and codeine, is refined. Heroin is produced from morphine by a simple chemical process, somewhat similar to the way LSD is produced from ergine.

Opium and the opiates, such as morphine, are useful in the treatment of severe pain. However, use of these drugs can

***Chinese smoking opium in an opium den, a practice that existed as long ago as the 12th century BC. When the Chinese entered the United States in the 1850s and 1860s, they brought with them the practice of opium smoking.***

lead to physiological dependence, and withdrawal from them causes symptoms such as chills, sweating, delusions, nausea, and diarrhoea. Addicts fear these symptoms and will go to any length to obtain more of the drug and thus avoid withdrawal. The use of opium derivatives, especially heroin, has also led to crime and social disruption.

The use of opium and its derivatives is illegal in many countries. In the United States, legislation to strictly control these drugs took many years to appear. The Pure Food and Drug Act of 1906 merely required manufacturers to state when opiates were among the ingredients in their medicines. Educational campaigns informed people of the dangers of patent medicines containing these drugs. The Harrison Narcotic Act of 1914 was passed to bring the United States into line with other countries that had already banned opiates. This had the effect of removing all legal supplies of opiates from the country, which forced users to buy drugs on the black market. Despite a 1924 law that banned the importation of heroin, many morphine addicts began to use this drug. Today, huge amounts of money are involved in black market sales, and many crimes are committed to obtain the money needed for heroin. And yet, the enactment of laws and the government's attempt to enforce them have been

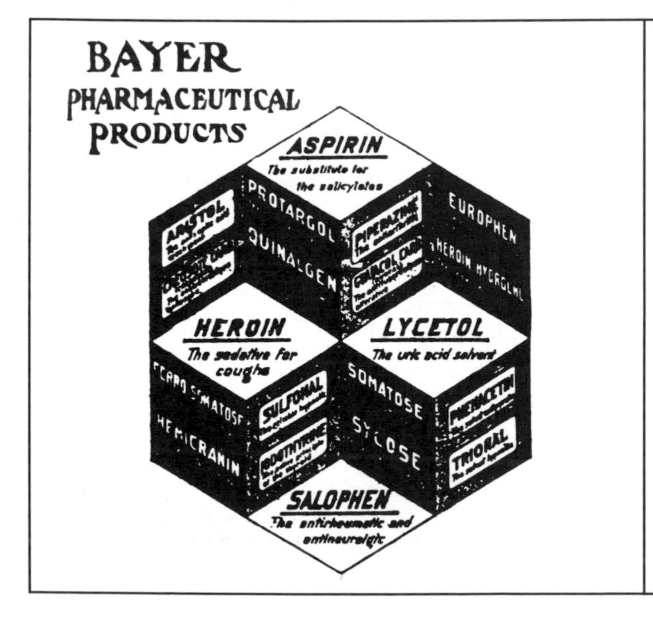


***An early newspaper advertisement for Bayer pharmaceutical products mentions heroin as a cough suppressant. The Harrison Narcotic Act of 1914 removed all legal supplies of opiates from the United States, and a 1924 law banned the importation of heroin.***

unable to decrease the number of addicts, which continues to grow.

In Britain, opiate addiction is treated in a very different way. Addiction to opium or its derivatives is treated as an illness rather than as a crime. Each addict is registered with the government and receives a small daily supply of the drug for personal use. Despite this, Britain now has a substantial drug problem although there are some signs that it may have reached a peak—at least with respect to heroin.

### *The Coca Plant*

The coca plant, *Erythroxylon coca,* is a member of the Erythroxylaceae family. Native to the Andes mountains of South America, coca is the source of the euphoric alkaloid called cocaine. The Incas regarded coca as a sacred plant, and by the 14th century the plant was widely used in northern South America. The Indians of South America (many of whom still use coca) mix the dry leaves with lime, place the mixture between the teeth and the cheek, and slowly suck the quid (a wad of something to be chewed). Used in this way, coca relieves fatigue and hunger pains.

During the hundreds of years of its use, coca has not

***A Kapi Indian gathering coca leaves. Native to the Andes Mountains, the coca plant,* Erythroxylon coca, *is the source of the stimulant and anaesthetic drug called cocaine.***

caused any social problems among the Indians of South America. It was the isolation of pure cocaine by Europeans in 1862 that led to the abuse of the plant. The famous psychologist Sigmund Freud experimented with cocaine and discovered that it relieved depression. Cocaine was later used in many patent medicines. At one time even the trademarked drink Coca-Cola contained cocaine. Cocaine usually does not cause a physiological addiction, such as that produced by heroin, but it does cause a psychological addiction. Cocaine users often suffer from depression after the initial euphoric feeling has passed.

Cocaine can be consumed in several forms. It can be taken as a snuff or dissolved in wine. The coca leaves can be mixed with tobacco for smoking, or the cocaine can be converted into what is called *free-base cocaine* and smoked. Of all these methods, the last one is the most dangerous. It involves the use of flammable chemical solvents and many people have been critically burned while attempting to smoke free-base cocaine.

### *Khat*

This is the dried leaves of an evergreen shrub, *Catha edulis,* of the family Celastraceae. Khat contains the alkaloids, cathine and cathinone, which are Controlled Drugs, although khat itself is not controlled.

Khat is cultivated in Ethiopia, Kenya, Djibouti and the Yemen Arab Republic where the khat habit is socially sanctioned in a deep cultural tradition. People meet together in small groups to chew the fresh leaves. The effects are similar to those of amphetamine, with feelings of elation, increased energy, alertness and self-esteem and an improved ability to communicate. However, psychological dependence can ensue and the habit can be debilitating.

### *Hallucinogenic Plants Today*

Thousands of years ago, while searching for new sources of food, people discovered hallucinogenic plants. Intrigued and frightened by the plants' awesome powers, the people gradually integrated the many plant substances into their customs and rituals: exorcisms, rites of passage, healing and divinitory rituals, and religious and mystical practices. In the course of the adoption of the use of hallucinogenic plants, countless numbers of people suffered and died from the effects of the psychoactive ingredients before strict rules governing their consumption were developed. Only then, within the highly ritualized environment, were the users somewhat protected from the often fatal effects of the drugs.

These same hallucinogens, with their wide array of effects, are available today. Even a small dose may produce a profound change in the central nervous system, the extent of which varies according to each individual's body. Because today the use of psychoactive plants is not integrated into society or controlled by elaborate rituals, it is the responsibility of the individual to learn about the effects of these potentially dangerous substances.

***A Colombian man burning leaves of*** **Cecropia sciadophila** ***to produce an alkaline ash that is mixed with powdered coca.***

## *Some Useful Addresses*

*In the United Kingdom:*

Advisory Council on the Misuse of Drugs
c/o Home Office, Queen Anne's Gate, London SW1H 9AT.

British Association for Counselling
87a Sheep Street, Rugby, Warwicks CV21 3BX.

Department of Education and Science
Elizabeth House, York Road, London SE1 7PH.

Health Education Council
78 New Oxford Street, London WC1A 1AH.

Home Office Drugs Branch
Queen Anne's Gate, London SW1H 9AT.

Institute for the Study of Drug Dependence
1–4 Hatton Place, Hatton Garden, London EC1N 8ND.

Medical Research Council
20 Park Crescent, London W1N 4AL.

Narcotics Anonymous
PO Box 246, c/o 47 Milman Street, London SW10.

National Association of Young People's
Counselling and Advisory Services
17–23 Albion Street, Leicester LE1 6GD.

Northern Ireland Department of Health and Social Services
Upper Newtownwards Road, Belfast BT4 3SF.

Release
1 Elgin Avenue, London W9.

Scottish Health Education Unit
21 Lansdowne Crescent, Edinburgh EH12 5EH.

Scottish Home and Health Department
St Andrews House, Edinburgh EH1 3DE.

Standing Conference on Drug Abuse
1–4 Hatton Place, Hatton Garden, London EC1N 8ND.

Teachers Advisory Council on Alcohol and Drug Education
2 Mount Street, Manchester M2 5NG.

*In Australia:*

Department of Health
PO Box 100, Wooden ACT, Australia 2606.

*In New Zealand:*

Drug Advisory Committee
Department of Health, PO Box 5013, Wellington.

Drug Dependence
11–23 Sturdee Street, Wellington.

Drug Dependency Clinic
393 Great North Road, Grey Lynn, Auckland.

Medical Services and Drug Control
Department of Health, PO Box 5013, Wellington.

National Drug Intelligence Bureau
Police Department, Private Bag, Wellington.

*In South Africa:*

South African National Council on Alcoholism and Drug Dependence (SANCA)
National Office, PO Box 10134, Johannesburg 2000.

A number of organizations in South Africa provide information and services in the field of drug dependence. SANCA will supply information on these, as will the government's Department of Health and Welfare.

## *Further Reading*


Abel, E.L. *marijuana.* New York: Plenum Press, 1980.

Brecher, Edward M., and the editors of *Consumer Reports. Licit and Illicit Drugs, The Consumer Union Report on Narcotics, Stimulants, Depressants, Inhalants, Hallucinogens, and Marijuana—including Caffeine, Nicotine, & Alcohol.* Mount Vernon, New York: Consumers Union, 1972.

De Rios, M.D. *Hallucinogens: Cross-Cultural Perspectives.* Albuquerque: University of New Mexico Press, 1984.

Emboden, W.A. *Narcotic Plants.* New York: Macmillan, 1972.

Furst, P.T. *Hallucinogens and Culture.* Novato, California: Chandler and Sharp, 1976.

Helmer, J. *Drugs and Minority Oppression.* New York: Seabury Press, 1975.

Rubin, V., ed. *Cannabis and Culture.* The Hague: Mouton Publishers, 1975.

Schultes, R.E. *Hallucinogenic Plants.* New York: Golden Press, 1976.

Schultes, R.E., and Hofmann, A. *Plants of the Gods.* New York: McGraw-Hill, 1979.

Schultes, R.E., and Hofmann, A. *The Botany and Chemistry of Hallucinogens* (Second Edition). Springfield, Illinois: Thomas, 1980.

Weil, A. and Rosen, W. *Chocolate to Morphine: Understanding Mind-Active Drugs.* Boston: Houghton Mifflin, 1983.

## *Glossary*

*absorption* the passage of a substance through a surface of the body into tissues and body fluids

*acupuncture* a practice of puncturing the body with long thin needles in order to relieve pain or cure disease

*addiction* a condition caused by repeated drug use, characterized by a compulsive urge to continue using the drug, a tendency to increase the dosage, and physiological and/or psychological dependence

*alkaloid* any chemical containing nitrogen, carbon, hydrogen, and oxygen, usually occurring in plants

*analgesic* a drug that produces an insensitivity to pain without loss of consciousness

*anaesthetic* a drug that produces loss of sensation, sometimes with loss of consciousness

*angiosperms* the flowering plant

*aphrodisiac* an agent that stimulates sexual desire

*aromatic ethers* a group of chemicals believed to contain hallucinogenic constituents

*atom* the smallest particle of an element that can exist either alone or in combination

*atropine* a psychoactive drug that has hallucinogenic properties at high doses

*ayahuasca* literally "vine of the souls", a hallucinogenic drink made from the bark of *Banisteriopsis* or *Tetrapteris,* lianas found mainly in South America

*bufotenin* a hallucinogenic compound that can be isolated from the plant genus *Anadenanthera*

*cannabinoids* the hallucinogenic compounds found in the resin of the plant genus *Cannabis*

*curandero* an American Indian shaman who specializes in curing

*cytisine* a quinolizidine alkaloid; probably the hallucinogen found in the plant species *Cytisus canariensis.* Cytisine can be toxic, causing nausea, convulsions, and even death

*dioecious* plants that are male or female

*DMT* N,N-dimethyltryptamine; a hallucinogenic compound found in the plant genus *Virola*

*dysentry* an intestinal disorder characterized by inflammation of the mucous membrane

*ergine* d-lysergic acid amide; the major hallucinogenic compound in the plant family Convolvulaceae

*ergot* a fungus, *Claviceps purpurea,* that infects rye. Ergot contains two types of compounds: one can cause gangrene poisoning; the other one is hallucinogenic

*euphoria* a mental high characterized by a sense of well-being

*extract* an alcoholic solution of the essential constituents of a plant

*ferment* to decompose

*freebase* the cocaine alkaloid, or base, that results when the hydrochloride is removed from cocaine hydrochloride; freebasing refers to the process by which the hydrochloride is freed

*glaucoma* a disease of the eye marked by increased pressure within the eyeball that can result in damage to the optic disk and a gradual loss of vision

*gout* a disease marked by a painful inflammation of the joints, deposits of urates in and around the joints, and usually an excessive amount of uric acid in the blood

*hallucinogen* a drug that produces a sensory impression that has no basis in external stimulation

*harmala alkaloids* the group of compounds that includes hallucinogens present in the drink *ayahuasca*

*heroin* a semisynthetic opiate produced by a chemical modification of morphine

*ibogaine* the main hallucinogenic ingredient in the plant species *Tabernanthe iboga*

*indole ring* a chemical structure consisting of eight carbon atoms and one nitrogen atom combined to form two adjacent rings. Chemical compounds that contain the indole ring are hallucinogenic

*jaundice* a condition in which the skin gains a yellowish colour because of the buildup within it of bile pigment

*jurema* a hallucinogenic drink once prepared and drunk by Indian tribes in eastern Brazil. Because these tribes are now extinct, the method for preparing *jurema* will never be known

*LSD* lysergic acid diethylamide; a hallucinogen derived

from a fungus that grows on rye or from morning-glory seeds

*mesa* a collection of Christian and satanic artifacts used by *curanderos* during a curing session

*mescaline* a hallucinogenic drug found in certain cacti

*morphine* the principal psychoactive ingredient of opium; this alkaloid produces sleep or a state of stupor, and is used as the standard against which all morphine-like drugs are compared

*narcotic* originally, a group of drugs producing effects similar to morphine; often used to refer to any substance that sedates, has a depressive effect, and/or causes dependence

*nutmeg* an aromatic seed, from the Myristicaceae family, that is used as a spice

*ololuiqui* a plant that contains a drug smilar to LSD

*opiate* a compound from the milky juice of the poppy plant *Papaver somniferum,* including opium, morphine, codeine, and their derivatives, such as heroin

*organic* derived from a living organism and containing carbon and hydrogen

*paraphernalia* the equipment and material used to administer or store illicit drugs

*peyote* a cactus that contains mescaline, a hallucinogenic drug, and is used legally by certain American Indians for religious and medical purposes

*physical dependence* an adaption of the body to the presence of a drug, such that its absence produces withdrawal symptoms

*psychoactive* altering mood and/or behaviour

*psychological dependence* a condition in which the drug user craves a drug to maintain a sense of well-being and feels discomfort when deprived of it

*quinolizidine alkaloid* a type of compound, found in the hallucinogenic plants *Cytisus canariensis* and *Sophora secundiflora*

*rheumatism* any of various conditions characterized by inflammation of muscles and joints

*San Pedro Trichocereus pachanoi;* a mescaline-containing cactus plant, grown mainly in the Andes mountains, from which the *cimora* drink is made

*serotonin* a neurotransmitter thought to be involved in neural mechanisms important in sleep and sensory perception

*sesquiterpenes* bacteria-inhibiting compounds found in the resin of the plant species *Humulus lupulus*

*sinsemilla* literally, "without seeds", refers to the process whereby the potent flowering top of the female plant is not pollinated and is therefore seedless

*solanine* a toxic alkaloid that, if ingested, can cause symptoms such as headaches, vomiting, diarrhoea, fever, apathy, restlessness, confusion, and hallucinations

*solvent* a liquid containing a solution of a substance

*sparteine* a quinolizidine alkaloid thought to produce hallucinations. Like cytisine, sparteine can cause nausea, convulsions, and can even be fatal

*synesthesia* confusion of the senses; for example, a sound may produce a mental image

*synthetic* artificially prepared; substances that are produced in a laboratory

*teratogenic* tending to cause development malformations

*toloache* a hallucinogenic hot water extract of the roots of the species *Datura inoxia*

*toxic* temporarily or permanently damaging to cells or organ systems of the body

*tropane* hallucinogenic compounds found in some species of *Atropa, Hyoscyamus, Mandragora,* and *Datura*

*tryptamine* a compound, sometimes hallucinogenic, found in the plant species *Anadenanthera peregrina* and *Anadenanthera colubrina*

*Virola* a plant genus, containing about 60 species, found mainly in parts of Central America and southern Brazil. About one-quarter of the species in the *virola* genus contain hallucinogenic substances

*withdrawal* the physiological and phychological effects that occur after the use of a drug is discontinued

*yohimbe* a compound found in the plant *Apocynum quebrancho-blanco,* which, when brewed as a tea and drunk, can cause hallucinations and an increase in blood pressure

*yopo* American Indian word for *Anadenanthera peregrina,* a powder that, when sniffed, causes nausea and hallucinations

# *Index*

***P. Mick Richardson,*** Ph.D., received his degree at the Royal Botanic Gardens, Kew and continued his research at the University of Reading and the University of Illinois at Urbana-Champaign. He has been published in several scientific journals and books and is a member of many scientific societies. Interested mainly in the secondary metabolites produced by plants, Dr P. Mick Richardson is an associate scientist at the New York Botanical Garden.

***Solomon H. Snyder,*** M.D., is Distinguished Service Professor of Neuroscience, Pharmacology and Psychiatry at The Johns Hopkins University School of Medicine. He has served as president of the Society for Neuroscience and in 1978 received the Albert Laster Award in Medical Research. He is the author of *Uses of Marijuana, Madness and the Brain, The Troubled Mind, Biological Aspects of Mental Disorder,* and edited *Perspective in Neuropharmacology: A Tribute to Julius Axelrod.* Professor Snyder was a research associate with Dr Axelrod at the National Institute of Health.

***Malcom Lader,*** D.Sc., Ph.D., M.D., F.R.C. Psych is Professor of Clinical Psychopharmacology at the Institute of Psychiatry, University of London and Honorary Consultant to the Bethlem Royal and Maudsley Hospitals. He is a member of the External Scientific Staff of the Medical Research Council. He has researched extensively into the actions of drugs used to treat psychiatric illnesses and symptoms, in particular the tranquillizers. He has written several books and over 300 scientific articles. Professor Lader is a member of several governmental advisory committees concerned with drugs.